# Atmospheric Halos

American Geophysical Union

# ANTARCTIC RESEARCH SERIES

**Physical Sciences**

American Geophysical Union

# ANTARCTIC RESEARCH SERIES

**Biological and Life Sciences**

BIOLOGY OF THE ANTARCTIC SEAS
    Milton O. Lee, *Editor*
BIOLOGY OF THE ANTARCTIC SEAS II
    George A. Llano, *Editor*
BIOLOGY OF THE ANTARCTIC SEAS III
    George A. Llano and Waldo L. Schmitt, *Editors*
BIOLOGY OF THE ANTARCTIC SEAS IV
    George A. Llano and I. Eugene Wallen, *Editors*
BIOLOGY OF THE ANTARCTIC SEAS V
    David L. Pawson, *Editor*
BIOLOGY OF THE ANTARCTIC SEAS VI
    David L. Pawson, *Editor*
BIOLOGY OF THE ANTARCTIC SEAS VII
    David L. Pawson, *Editor*
BIOLOGY OF THE ANTARCTIC SEAS VIII
    David L. Pawson and Louis S. Kornicker, *Editors*
BIOLOGY OF THE ANTARCTIC SEAS IX
    Louis S. Kornicker, *Editor*
BIOLOGY OF THE ANTARCTIC SEAS X
    Louis S. Kornicker, *Editor*
BIOLOGY OF THE ANTARCTIC SEAS XI
    Louis S. Kornicker, *Editor*
BIOLOGY OF THE ANTARCTIC SEAS XII
    David L. Pawson, *Editor*
BIOLOGY OF THE ANTARCTIC SEAS XIII
    Louis S. Kornicker, *Editor*
BIOLOGY OF THE ANTARCTIC SEAS XIV
    Louis S. Kornicker, *Editor*
BIOLOGY OF THE ANTARCTIC SEAS XV
    Louis S. Kornicker, *Editor*
BIOLOGY OF THE ANTARCTIC SEAS XVI
    Louis S. Kornicker, *Editor*
BIOLOGY OF THE ANTARCTIC SEAS XVII
    Louis S. Kornicker, *Editor*
BIOLOGY OF THE ANTARCTIC SEAS XVIII
    Louis S. Kornicker, *Editor*
BIOLOGY OF THE ANTARCTIC SEAS XIX
    Louis S. Kornicker, *Editor*
BIOLOGY OF THE ANTARCTIC SEAS XX
    Louis S. Kornicker, *Editor*
BIOLOGY OF THE ANTARCTIC SEAS XXᵀ
    Louis S. Kornicker, *Editor*
BIOLOGY OF THE ANTARCTIC SEAS XXII
    Stephen D. Cairns, *Editor*

ANTARCTIC TERRESTRIAL BIOLOGY
    George A. Llano, *Editor*
TERRESTRIAL BIOLOGY II
    Bruce Parker, *Editor*
TERRESTRIAL BIOLOGY III
    Bruce Parker, *Editor*


ANTARCTIC ASCIDIACEA
    Patricia Kott
ANTARCTIC BIRD STUDIES
    Oliver L. Austin, Jr., *Editor*
ANTARCTIC PINNIPEDIA
    William Henry Burt, *Editor*
ANTARCTIC CIRRIPEDIA
    William A. Newman and Arnold Ross
BIRDS OF THE ANTARCTIC AND SUB-ANTARCTIC
    George E. Watson
ENTOMOLOGY OF ANTARCTICA
    J. Linsley Gressitt, *Editor*
HUMAN ADAPTABILITY TO ANTARCTIC CONDITIONS
    E. K. Eric Gunderson, *Editor*
POLYCHAETA ERRANTIA OF ANTARCTICA
    Olga Hartman
POLYCHAETA MYZOSTOMIDAE AND SEDENTIARIA OF ANTARCTICA
    Olga Hartman
RECENT ANTARCTIC AND SUBANTARCTIC BRACHIOPODS
    Merrill W. Foster
ANTARCTIC AND SUBANTARCTIC PYCNOGONIDA: AMMOTHEIDAE AND AUSTRODECIDAE
    Stephen D. Cairns, *Editor*

Volume 64

ANTARCTIC
RESEARCH
SERIES

# Atmospheric Halos

# by Walter Tape

CELEBRATING 75 YEARS OF LEADERSHIP
75
1919–1994
AGU

American Geophysical Union
Washington, D.C.
1994

*Volume 64* | ANTARCTIC
RESEARCH
SERIES

Front cover: Halo display, South Pole, January 2, 1990.

Back cover: Computer simulation of the display shown on the front cover.

**Library of Congress Cataloging-in-Publication Data**

Tape, Walter.
   Atmospheric halos / by Walter Tape.
      p.     cm. — (Antarctic research series ; v. 64)
   Includes bibliographical references and index.
   ISBN 0-87590-834-9
   1. Halos (Meteorology)     I. Title. II. Series.
QC976.H15T36   1994
551.5'67—dc20                                                                    93-29785
                                                                                     CIP

ISBN 0-87590-834-9
ISSN 0066-4634

Published by
American Geophysical Union
With the aid of grant DPP-89-15494 from the
National Science Foundation

Printed in the United States of America.

CONTENTS

# The Antarctic Research Series:

# STATEMENT OF OBJECTIVES

The Antarctic Research Series provides for the presentation of detailed scientific research results from Antarctica, particularly the results of the United States Antarctic Research Program, including monographs and long manuscripts.

The series is designed to make the results of Antarctic fieldwork available. The Antarctic Research Series encourages the collection of papers on specific geographic areas within Antarctica. In addition, many volumes focus on particular disciplines, including marine biology, oceanology, meteorology, upper atmosphere physics, terrestrial biology, geology, glaciology, human adaptability, engineering, and environmental protection.

Topical volumes in the series normally are devoted to papers in one or two disciplines. Multidisciplinary volumes, initiated in 1990 to enable more rapid publication, are open to papers from any discipline. The series can accommodate long manuscripts and utilize special formats, such as maps.

Priorities for publication are set by the Board of Associate Editors. Preference is given to research manuscripts from projects funded by U.S. agencies. Because the series serves to emphasize the U.S. Antarctic Research Program, it also performs a function similar to expedition reports of many other countries with national Antarctic research programs.

The standards of scientific excellence expected for the series are maintained by the review criteria established for the AGU publications program. Each paper is critically reviewed by two or more expert referees. A member of the Board of Associate Editors may serve as editor of a volume, or another person may be appointed. The Board works with the individual editors of each volume and with the AGU staff to assure that the objectives of the series are met, that the best possible papers are presented, and that publication is timely.

Proposals for volumes or papers offered should be sent to the Board of Associate Editors, Antarctic Research Series, at 2000 Florida Avenue, N.W., Washington, D.C. 20009. Publication of the series is partially supported by a grant from the National Science Foundation.

Board of Associate Editors
*Antarctic Research Series*

# PREFACE

*The moon, little past her full, had a great ring around her, faintly prismatic; and equidistant from her, where a line through her centre parallel with the horizon would cut the ring, were two other moons distinct and clear. It was a strangely beautiful thing, this sight of three moons sailing aloft through the starry sky, as though the beholder had been suddenly translated to some planet that enjoys a plurality of satellites...*

Archdeacon Hudson Stuck
Bettles, Alaska, 1906
*Ten Thousand Miles with a Dog Sled*

Archdeacon Stuck was seeing halos formed in moonlight. Their daytime counterparts are surprisingly common, not only in the arctic but in temperate climates as well. Many halos are possible, forming arcs of colored or white light almost anywhere in the sky. Their occasional brilliance, variety, and exotic shapes have impressed skywatchers for centuries.

This book introduces halos and tries to convey some of their beauty. It tells much of what is known about them: how they arise, how so many are possible, and why some are rare, while others occur every few days or so. No such insights, of course, are necessary for the enjoyment of a halo display; indeed, an elaborate display is one of Nature's wonders. Nevertheless, understanding can add to enjoyment, especially since the makings of a great display turn out to be as remarkable as the display itself.

Many scientists have been fascinated by halos. M. A. Bravais in the nineteenth century, followed by Alfred Wegener in the early twentieth, pioneered the subject. Since about 1970, however, three developments have brought dramatic progress. First, good halo photographs have begun to replace the classic, but not always reliable, halo drawings in the literature. (Anyone who sees a rare halo can obtain useful photographs, as explained in Appendix A.) Second, atmospheric ice crystals — the cause of halos — have been collected and photographed during halo displays, thereby clarifying the relation between ice crystals and halos. Third, computers have been introduced with spectacular results. Previously unwieldy calculations are now performed rapidly, and the resulting predictions are displayed in halo simulations that can be readily compared with halo photographs.

This book combines halo photographs, ice crystal samples, and computer simulations to tell what is known about halos. It relies on as little else as possible; the approach, wherever possible, is: What can be learned from the halo photographs, crystal samples, and simulations alone? Thus, some relevant topics have been omitted: cloud physics, for example, which might explain why the crystals are shaped the way they are, and fluid dynamics, which might explain why the crystals fall the way they do.

The book, then, is organized around actual halo displays. The displays were selected from hundreds of halo and ice crystal observations made mostly in Wisconsin, Alaska, and Antarctica. Those selected were chosen because they were typical, or because they were spectacular, or because they were relatively uncomplicated.

Thanks in part to visits to the Antarctic interior, where halos can be exceptional, the halo photographs are outstanding. Several show halos never before photographed, and one, Figure 3-2, shows a halo that is essentially new. Thus the photographs are at the heart of the book. But a photograph that compresses the entire sky onto a piece of paper can only hint at the beauty and scope of a great display. Also, the intensity variations among halos in a slide make it difficult to reproduce all of them in a single print. In short, photographs cannot replace skywatching. If you are not already aware of halos, I hope that browsing through the book will induce you to watch for these lovely phenomena.

The book is meant for conscientious readers with no prior knowledge of halos. Most calculations, including the theoretical derivations of halo shapes, have been left to the computer. Other technical material has been largely relegated to appendices and notes. Where this was impossible, the text gives warning.

Chapters 1, 2, 4, 5, and 6 will answer most questions of the casual halo watcher. Because there are many halos, I suggest reading at a leisurely pace. A satisfying, though slow, initiation to the book is to watch the sky regularly and then to refer to the book when an unfamiliar halo appears. In most localities halos can occur at any time. Other than sunglasses to diminish the glare of the sky, no special equipment is needed to observe and enjoy them.

# ACKNOWLEDGEMENTS

Naturally I am indebted to previous contributors to halo theory, but I owe special thanks to those who pioneered computer simulations of halos: Robert Greenler and his colleagues at the University of Wisconsin - Milwaukee, and E. Tränkle and F. Pattloch at Freie Universität Berlin.

I want to thank Marty Getz, Robert Greenler, Günther Können, and Gerald Tape, who read early drafts of the book and made helpful suggestions. I also want to thank Ming-Ying Wei at the AGU Books Board and Bernhard Lettau, John Lynch, and Ronald C. Taylor at the National Science Foundation, all of whom helped to make this book a reality.

I was fortunate to spend four summer seasons in Antarctica studying halos and ice crystals. Three were at Amundsen - Scott South Pole Station, and one at the Soviet Station Vostok. I want to thank my hosts at both stations for their support and hospitality. For two of those seasons I collaborated with Günther Können, who made the work more productive and more fun.

This work was supported by National Science Foundation grants DPP-8314178 and DPP-8816515, and by the University of Alaska Fairbanks. The color reproductions were supported by the National Science Foundation through award OPP-9347911 representing contributions from the Polar Oceans and Climate Systems Program of the Office of Polar Programs, and the Physical Meteorology Program, Division of Atmospheric Sciences.

**Figure 1-1.** Halo formed in ice crystals between the photographer and the building. Fairbanks, Alaska, February 17, 1983.

CHAPTER 1

## HALOS FROM PLATE CRYSTALS

Halos — arcs or spots of light in the sky — are caused by the play of sunlight on ice crystals in the atmosphere. That may sound preposterous to inhabitants of temperate climates, where halos are common but where the responsible ice crystals are normally high in the atmosphere, out of sight and out of mind. But in cold climates the crystals sometimes occur near the ground, sparkling in the sunlight as they fall. Halos may then appear in front of buildings or other objects, and the association of halos with ice crystals becomes obvious.

During these ground level halo displays, the ice crystals can be collected and examined with a microscope. Smaller than the familiar stellar snowflakes, the crystals usually turn out to be tiny hexagonal prisms, either platelike or columnar. Comparing the crystals with the halos reveals that the two types of crystals — plates and columns — tend to cause different types of halos. The halos in Chapter 1 are halos that arise in plate crystals.

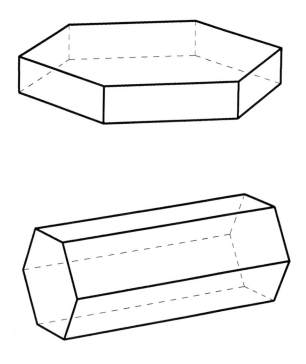

**Figure 1-2.** (Top) Plate crystal. (Bottom) Column crystal.

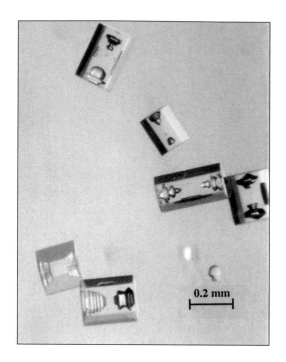

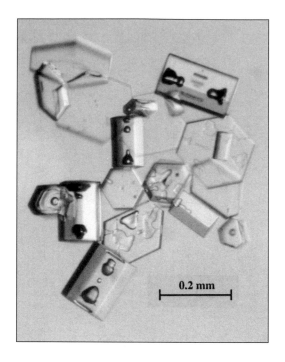

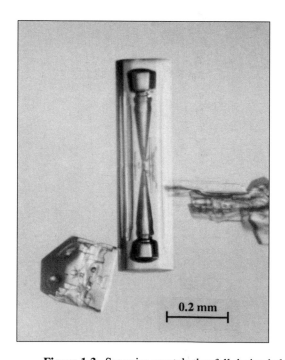

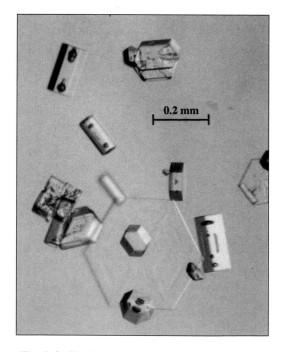

**Figure 1-3.** Some ice crystals that fell during halo displays. (Top left) Stubby columns. (Top right) Columns and plates. (Bottom left) Large column with beautiful internal structure. (Bottom right) Columns and plates.

### Display 1-1
### South Pole, January 2, 1990

**Parhelia and circumzenith arc.** The most common halos due to plate crystals are the parhelia and the circumzenith arc. In the halo display shown below, the parhelia, or sundogs, are the two bright spots on either side of the sun, and the circumzenith arc is the colored arc near the top.

These halos are common indeed, though normally not as bright as here. In my experience in Wisconsin, the parhelia occur about 60 days a year. If you fail to see them, you may be looking too close to the sun. Their angular distance from it is about 22°, a separation that can be approximated by extending your arm, spreading your fingers, and placing your thumb on the sun; your little finger should then be on the parhelion. The circumzenith arc appears about 25 days a year, with spectral colors sometimes as vivid as a rainbow. As its name suggests, it lies on a circle centered at the zenith. Most people never see it, because they do not look up.

The ice crystals shown in Figure 1-5 are plate crystals. They were collected as they fell during the halo display. No other types of crystals were seen, so crystals like these must have given rise to the parhelia and the circumzenith arc.

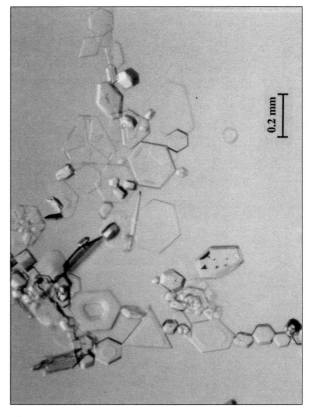

**Figure 1-4.** Parhelia at left and right; circumzenith arc at top. The sun is obscured by the tower. The wide angle lens that was used compresses the display and makes the circumzenith arc look too low. (Display 1-1)

**Figure 1-5.** Some crystals collected during Display 1-1. Nearly all are plates. The crystals are collected in hexane, and if the container is moved, they sometimes bunch together, as here.

**The orientations of the crystals.** Computer simulations can display the halos theoretically caused by sunlight falling on ice crystals having specified shapes and orientations. The simulation in Figure 1-6, for example, shows parhelia and a circumzenith arc theoretically caused by plate crystals falling with roughly horizontal orientations. It is not obvious that crystals should fall this way; in fact, nineteenth century scientists thought plate crystals would orient vertically, knifing through the air as they fell. But the close resemblance between the simulation and the halo photograph indicates that at least in this display the crystals were indeed more or less horizontal.[1]

Plate crystals falling with nearly horizontal orientations are called oriented plates. It is oriented plates, then, that seem to be the cause of the parhelia and circumzenith arc. Many other halo and ice crystal observations, in conjunction with simulations, point to the same conclusion.

**How the simulation was made.** Halos are caused by the refraction (bending) and reflection of sunlight in ice crystals. Figure 1-7 shows a plate crystal and a light ray path through the crystal, as calculated and drawn by computer using the laws of refraction and reflection. An observer whose eye intercepts the outgoing ray from the crystal would see a point of light in the direction opposite to that of the ray. The computer plotted a tiny dot at that point in the simulation. Then the process was repeated: the computer considered another crystal and light ray from the sun, calculated the path of the ray through the crystal, and plotted the dot determined by the outgoing ray.[2] By doing this for thousands of crystals having different shapes and orientations, the computer simulated the halo display; as more and more dots were plotted, halos began to appear as concentrations of dots.

The computer had to be told the shapes and orientations of the crystals to be used in the simulation. The shapes were modeled after the real crystals, and hence the computer was told to use plate crystals. But getting the orientations right was partly trial and error. Two populations of plate crystals were eventually chosen, one having tilts of 4°, and the other, 40°.[3] (The tilt of a plate crystal is the angle that the crystal makes with a horizontal plane. The smaller the tilts, the better — more nearly horizontal — are the orientations.) The second population, whose crystals are so poorly oriented that they might better not be called oriented plates at all, was introduced to make the faint circular halo in the simulation.

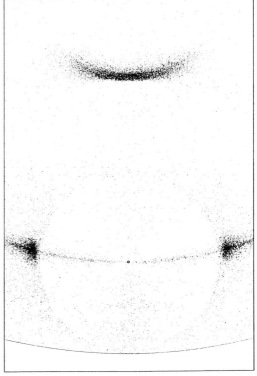

**Figure 1-6.** Simulation of Figure 1-4, showing halos theoretically caused by plate crystals oriented more or less horizontally.

The success of the simulation in matching the photograph shows that the crystal tilts in the real display were probably close to those in the simulation. Of course, it also indicates that sunlight falling on ice crystals is indeed the cause of halos.

**Halos and light ray paths.** Each dot in the simulation arises from a computed crystal and ray path through the crystal. The computer can draw the crystal and ray path giving rise to any specified dot. If dots are specified on different halos, their ray paths turn out to be different; different halos have different ray paths. For example, the ray path in Figure 1-7 makes the left parhelion, whereas the path in Figure 1-8 makes the circumzenith arc.

All of the crystal and ray path figures in this book have been drawn by computer. Each corresponds to a dot in a simulation and accurately portrays a potentially real ray path. For the common halos most ray paths have long been known, but for other halos the computer sometimes reveals complex and unforeseen ray paths.

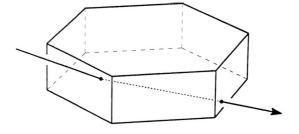

**Figure 1-7.** One possible light ray path through a crystal. This path contributes to the left parhelion. All of the crystals shown in the ray path figures of Chapter 1 are oriented plates, that is, plate crystals with nearly horizontal orientations.

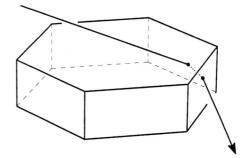

**Figure 1-8.** Ray path contributing to the circumzenith arc. The ray enters the top basal face of the crystal and exits one of the six prism faces, whereas for the parhelion ray path of Figure 1-7, the ray enters a prism face and exits an alternate (i.e., neither adjacent nor opposite) prism face.

---

1 What often goes unsaid is that simulations using other crystal orientations do not reproduce the halos successfully. For instance, simulations using vertical plate crystals make neither parhelia nor a circumzenith arc.

2 Actually, not all such dots were plotted. Rather, each dot was plotted with probability equal to the intensity of its outgoing ray. For example, of 1,000 dots associated with rays of intensity 0.2, about 200 would have been plotted. The density of dots in the simulation therefore indicates halo intensity. See Appendix F for a more complete description of the simulation program.

3 More precisely, in the first population the tilts were normally distributed with mean zero and standard deviation 4°. A table for the standard normal distribution shows that about 68% of the crystals would have been within 4° of horizontal, and 95% within $2 \times 4° = 8°$ of horizontal. Similarly for the second population. See step 3 of Appendix F.

### Display 1-2
### Eau Claire, Wisconsin, February 12, 1978

The photograph at the right shows another parhelion. No crystal sample accompanies this display, but I assume that the parhelion was caused by oriented plate crystals, just as in Display 1-1. The simulation shows the left parhelion that would be theoretically expected if the plates had fallen with tilts of 1.5°. The agreement between the simulation and the photograph is quite good.

**How the parhelia are formed.** The figure below shows top views of three crystals and ray paths that contributed to the parhelion in the simulation. These views reveal that the outgoing rays tend to accumulate in a single direction: The three incoming rays from the sun are parallel, and in spite of the quite different crystal orientations, the outgoing rays are also nearly parallel. A concentration of light — the parhelion — is therefore perceived in the direction opposite to that of the outgoing rays.

This concentrating of light rays is crucial to the existence of several halos, including the parhelia. If a disproportionate share of crystal orientations have their outgoing rays in nearly the same direction, then many different orientations seem to light the same small region of sky.

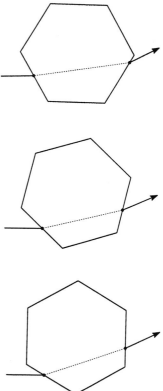

Homemade parhelia provide an illustration. To make a parhelion, place an equilateral triangular prism (or hexagonal prism, if available) in a bright beam of white light, and let the two outgoing colored beams from the prism fall upon a wall. With the prism axis vertical, spin the prism about its axis, thereby mimicking a large collection of oriented plates. (Attach the prism to a hand drill, for example.) The two moving colored spots on the wall, surprisingly, spend most of their time near just two points. The concentrations of light at these two points, both at the same height, are the "parhelia" (Figure 1-12). They are convincing facsimiles of real parhelia in the sky.

Of course, neither the computer simulation nor the spinning prism explains exactly why light accumulates in certain directions to form the parhelia. Except for some relatively simple halos, such as the parhelic circle and 120° parhelia of the next section, I have made no attempt in this book to explain conceptually why the various halos look the way they do.[4] Instead, the work has been left to the computer; by calculating ray paths through thousands of crystals having specified shapes and orientations, the computer predicts the appearance of the resulting halos.

**Figure 1-9.** Plate crystals and ray paths contributing to a left parhelion. Despite the different crystal orientations, the outgoing rays are nearly parallel.

---

4 Several books do attempt halo derivations, but the methods tend to be more computational than conceptual. See, for example, Visser [1942-1961] or Humphreys [1940].

**Figure 1-10.** Left parhelion. (Display 1-2)

**Figure 1-11.** Simulation of the display. Oriented plate crystals made the halos.

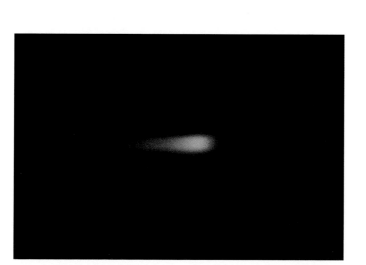

**Figure 1-12.** Artificial parhelion produced by a spinning glass prism in a light beam from a slide projector.

8

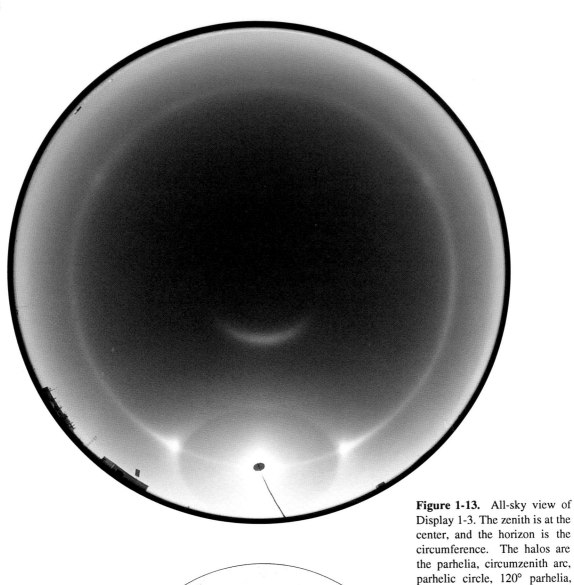

**Figure 1-13.** All-sky view of Display 1-3. The zenith is at the center, and the horizon is the circumference. The halos are the parhelia, circumzenith arc, parhelic circle, 120° parhelia, and 22° circular halo, here flattened by the fisheye lens. The 22° halo is discussed in Chapter 4.

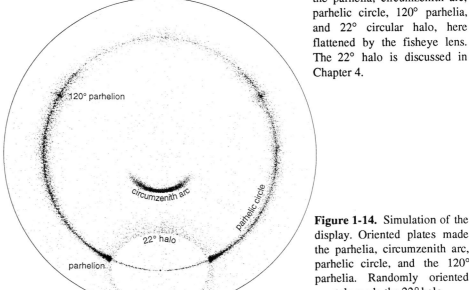

**Figure 1-14.** Simulation of the display. Oriented plates made the parhelia, circumzenith arc, parhelic circle, and the 120° parhelia. Randomly oriented crystals made the 22° halo.

## Display 1-3
## South Pole, January 4, 1985

This display strikingly supports the association of the parhelia and circumzenith arc with plate crystals. By the end of the display, virtually all the falling crystals were plates, and the parhelia and circumzenith arc were clear (Figures 1-15 and 1-16).

Any other halos then present must also have been caused by plates. There were two such halos, known as the parhelic circle and the 120° parhelia. Simulations show that to produce these halos, the plates must have oriented nearly horizontally. That is, like the parhelia and circumzenith arc, the new halos were due to oriented plates.

**Figure 1-15.** Some crystals collected during the display.

**Figure 1-16.** Late stage of Display 1-3. The 22° halo seen earlier has nearly disappeared. This photograph was taken while the crystals in Figure 1-15 were being collected.

**Parhelic circle.** In the all-sky photograph, Figure 1-13, the parhelic circle is the huge white halo parallel to the horizon and passing through the sun. When bright and complete like this one, the parhelic circle can be imposing. Unfortunately, good ones occur infrequently, probably less often than once a year in most localities. Weak fragments appear more often.

The figure below shows two ray paths for the parhelic circle. In each the key feature is a reflection, either external or internal, from the nearly vertical face of an oriented plate crystal. Reflections from vertical faces produce no change in the ray's angle of elevation, and in the lower crystal, entry and exit produce mutually offsetting changes. Hence these ray paths light points in the sky at the same elevation as the sun; that is, they light points on the parhelic circle.[5] The portion of the parhelic circle near the sun is lit by rays that meet the reflecting face nearly tangentially, since the reflection hardly changes the ray's direction. Rays more inclined to the reflecting face light portions farther from the sun.

In the lower crystal, the refraction at exit cancels the refraction at entry. The parhelic circle thus involves reflection but no net refraction.[6] Such halos are normally white, whereas halos that do involve net refraction, like the parhelia and circumzenith arc, usually show some color, because the amount of refraction depends on the color of the light.

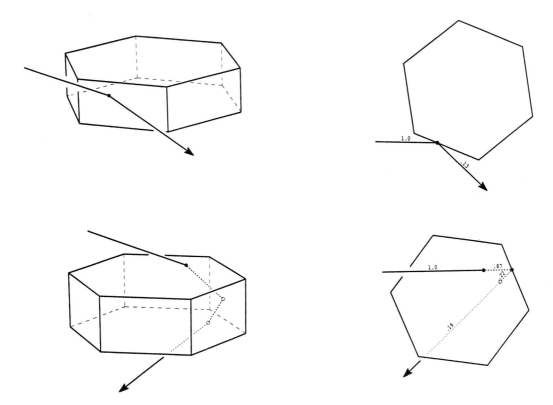

**Figure 1-17.** (Left) Two common ray paths for the parhelic circle. Open circles indicate encounters of the ray with crystal faces hidden from view; solid circles indicate encounters with visible faces. Thus in the lower diagram the ray enters the top basal face of the crystal, reflects internally from a prism face, and exits the bottom basal face. (Right) Same ray paths seen from above. The numbers give the intensities of the ray path segments. Each encounter of the ray with a face produced a reflected ray and a transmitted ray, only one of which is shown; the omitted rays account for the decreasing intensities.

**120° parhelia.** In the all-sky photograph the sun and two white spots on the parhelic circle form the vertices of a giant imaginary equilateral triangle. The white spots are the 120° parhelia.

The 120° parhelia are not common. Even when present, they are easily missed, especially if there is no parhelic circle to draw attention to them. But when the ordinary parhelia are bright, look far from the sun for the 120° parhelia.

The figure below shows a common ray path for the right 120° parhelion. The ray enters the top basal face of an oriented plate crystal, reflects internally from a prism face, reflects again from an adjacent prism face, and exits the bottom basal face. As shown in the right hand view, the pair of reflections increases the ray's azimuth by 120°, and entry and exit leave the azimuth unchanged. And as with the parhelic circle ray discussed above, the various crystal face encounters produce no net change in the ray's angle of elevation. So indeed, this ray must light a point on the parhelic circle 120° to the right of the sun, namely, the right 120° parhelion. The essential and remarkable feature of the ray path is that the azimuthal increase (120°) is independent of the crystal orientation, so long as the crystal remains horizontal.[7] Many different crystal orientations therefore light the same point in the sky.

Thin plates usually make poor 120° parhelia, even if they orient horizontally. If the crystal in the figure were somewhat thinner, the ray would probably exit the bottom face without reaching the second prism face; it would then contribute to the parhelic circle but not to the 120° parhelia. And if the crystal were thinner still, the ray would exit the bottom face without reaching even the first prism face. This helps to explain the infrequent occurrence of the 120° parhelia.

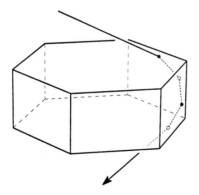 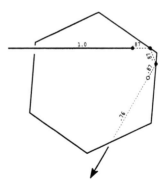

**Figure 1-18.** (Left) Common ray path for the right 120° parhelion. (Right) Same ray path seen from above. Here the internal reflections happen to be total; no intensity is lost.

---

5  I do not mean, of course, that the parhelic circle, or any other halo, is "lit" in the same sense that a spotlight illuminates an object. Two observers perceive the same halo as coming from two different regions of sky, but neither region is selectively "lit". Think of halos not as positions, but rather as directions; like rainbows in lawn sprinklers, halos seem to move when you do.

6  Some uncommon ray paths for the parhelic circle do give net refraction. See Table 1 of Appendix E.

7  A geometrical argument shows that the azimuthal deviation is 2 × (180 - A), where A is the angle between the two reflecting faces. Here A = 120° and the deviation is 2 × (180 - 120) = 120°. This simple yet surprising fact is independent of the angle of incidence at the first reflecting face.

## Display 1-4
## Castner Glacier, Alaska, October 23, 1982

A bright and colorful circumzenith arc is a beautiful sight. Here the circumzenith arc was exquisite. The parhelion had practically no vertical extent, indicating very small tilts for the plate crystals. Simulations would show the tilts to be about half a degree.

Other, faint halos are also discernible. They are not, however, due to plate crystals and will be discussed in Chapter 2.

**Figure 1-19.** Sharp parhelion and parhelic circle at right, and fine circumzenith arc. (Display 1-4)

# CHAPTER 2

## HALOS FROM COLUMN CRYSTALS

Chapter 1 looked at halos arising in oriented plates, that is, in plate crystals falling with their (principal) axes nearly vertical. Different halos turned out to be associated with different light ray paths through the crystals.

Chapter 2 looks at halos arising in column crystals falling with their axes nearly horizontal (but otherwise unconstrained). Such crystals are called singly oriented columns. The halo displays in Chapter 2 were chosen to show that singly oriented columns do occur and that, like oriented plates, they produce their own distinctive halos. Surprisingly many halos can arise in singly oriented columns, different halos again being associated with different ray paths. These halos can make stunning displays.

### Display 2-1
### South Pole, January 22, 1985

**Upper and lower tangent arcs.** This display was weak and the crystal swarm sparse. Yet the display is significant, because the crystals were almost all of one kind — columns. Column crystals must therefore have been the cause of the halos.

There were only two halos — the upper and lower tangent arcs, or, more simply, tangent arcs. (They would have been tangent to the 22° halo, had it been present.) Figure 2-2 shows the upper tangent arc and includes enough sky to confirm that the parhelia and the 22° halo were absent. Figure 2-3 shows the lower tangent arc, which was below the horizon and appears only as a concentration of sparkles against the hut. We know that column crystals were responsible for these halos, but what were their orientations?

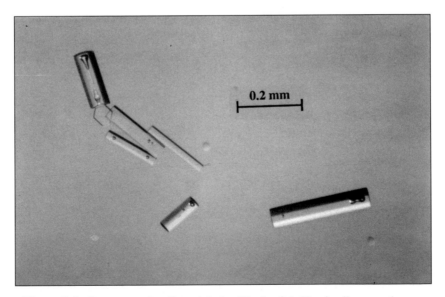

**Figure 2-1.** Some crystals collected during Display 2-1. Nearly all were columns.

**Figure 2-2.** Upper tangent arc. (Display 2-1)

**Figure 2-3.** Lower tangent arc seen in individual crystals between the photographer and the hut. (Display 2-1)

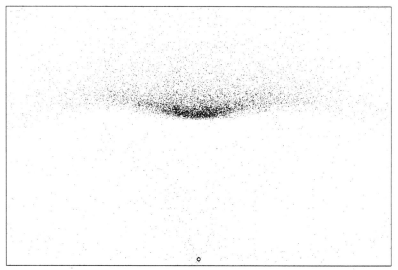

**Figure 2-4.** Simulation of Figure 2-2. Singly oriented columns, that is, column crystals with their axes nearly horizontal, made the halo — an upper tangent arc.

The simulation above was made with singly oriented columns. Its close resemblance to the photograph indicates that the crystals in the real display were also singly oriented. The upper tangent arc, then, was apparently due to singly oriented columns. A comparable simulation below the sun would show that singly oriented columns could cause the lower tangent arc as well.

Other halo and crystal observations, in conjunction with simulations, confirm that the tangent arcs are due to singly oriented columns. The photograph below, for example, shows crystals from a display nearly identical to Display 2-1. Except for an occasional small plate, all were imperfect large columns.

The tilt of a column crystal is the angle between the crystal axis and a horizontal plane. Singly oriented columns therefore have small tilts. The smaller the tilts, the better (more nearly horizontal) are the orientations and the better defined are the halos. The tilts of the crystals used in making the simulation above were 2.5°. That may sound small, but the resulting upper tangent arc is only poorly defined. Such mediocre upper tangent arcs are common; the one described next is unusual.

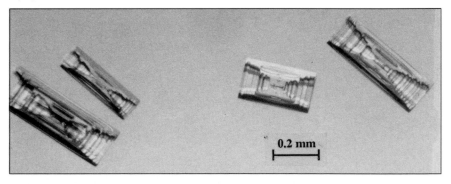

**Figure 2-5.** Some crystals collected during another halo display, similar to Display 2-1. South Pole, January 21, 1990.

## Display 2-2
## South Pole, January 19, 1985

If the tilts of the column crystals become smaller, the upper tangent arc becomes better defined, eventually displaying the beautiful and somewhat fantastic shape shown on the facing page.

The crystal sample collected during this display contained both plates and columns. From observations like those in Chapter 1, I assume that the plates oriented nearly horizontally and caused the parhelia. Then, as in Display 2-1, the columns must have caused the upper tangent arc. To do so, their tilts must have been small; to produce the upper tangent arc in the simulation, I had to give the columns tilts of only 0.6°.

As explained in Chapter 1, the computer can be asked to find ray paths responsible for lighting specified regions of sky in a simulation. The resulting ray paths for the tangent arcs are shown below. Like rays for the parhelia, they enter a prism face of a crystal and exit an alternate prism face.

To see how the tangent arcs form, imagine a sky full of horizontal column crystals with axes all perpendicular to the incoming sunlight. These hypothetical crystals would cause "parhelia" above and below the sun, rather than at left and right. If the axes were then turned so as to point in some other, single horizontal direction, the parhelia would shift. The tangent arcs are the juxtaposition of many such par-helia. They can be illustrated with the spinning prism described in Chapter 1.

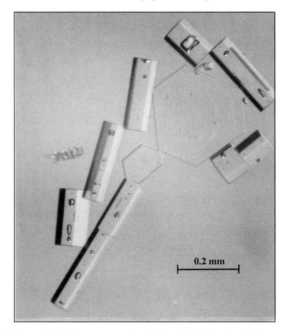

0.2 mm

**Figure 2-6.** Some crystals collected during the display.

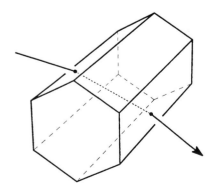
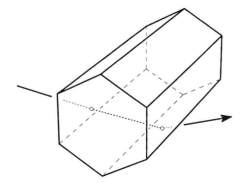

**Figure 2-7.** Ray paths for (left) the upper tangent arc and (right) the lower tangent arc. All of the crystals shown in the ray path figures of Chapter 2 are singly oriented columns.

**Figure 2-8.** Upper tangent arc and parhelia. (Display 2-2)

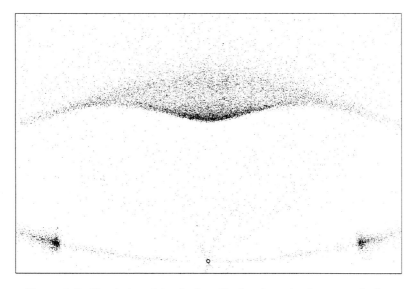

**Figure 2-9.** Simulation of the display. Singly oriented columns made the upper tangent arc; oriented plates made the parhelia.

## Display 2-3
## South Pole, January 2, 1990

**Supralateral and infralateral arcs.** When the upper tangent arc is strong, look further from the sun for the supralateral and infralateral arcs. On the facing page, the supralateral arc is the huge, nearly semicircular arc in the upper half of the photograph, and the infralateral arcs are the colorful arcs at the lower left and right. The wide angle photograph fails to convey their size; you must think big to see them in reality.

The crystal sample collected during this spectacular display contained primarily plates and short columns. Based partly on the sample, several populations of crystals were used in simulating the display (Figure 2-12). Only the largest population is relevant at the moment; it consisted of singly oriented columns and was responsible for the supralateral and infralateral arcs. It also made the the tangent arcs, part of the parhelic circle intensity, and some weaker halos not yet discussed.

A ray for the supralateral arc, shown in Figure 2-13, enters a prism face of a singly oriented column and exits a basal face. A ray for the infralateral arcs enters a basal face and exits a prism face.

As will be explained in Chapter 4, many 46° circular halos turn out on scrutiny to be supralateral and infralateral arcs. Even so, these arcs occur infrequently. One reason is that their ray paths, unlike those for the tangent arcs, require reasonably intact basal faces. The crystals of Figure 2-5, for example, because of the large cavities in their ends, could never produce supralateral and infralateral arcs.

**Parhelic circle.** Recall from Chapter 1 that the parhelic circle can be caused by reflections from the vertical prism faces of oriented plate crystals. Vertical faces, now basal faces, occur in singly oriented columns as well, and singly oriented columns can therefore cause the parhelic circle. Figure 2-15 shows two common ray paths.

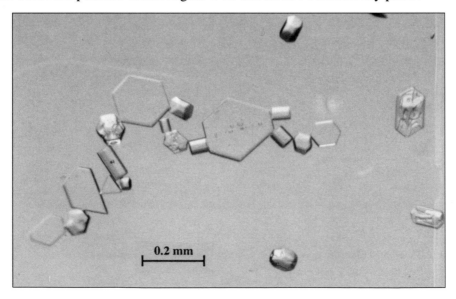

0.2 mm

**Figure 2-10.** Some crystals collected during the display.

**Figure 2-11.** Halo display with supralateral and infralateral arcs. (Display 2-3)

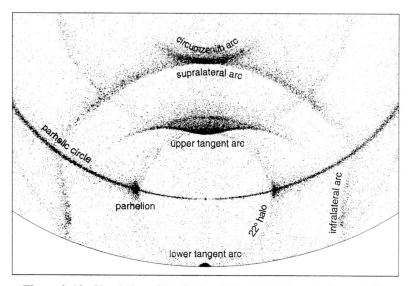

**Figure 2-12.** Simulation of the display. Oriented plates made the parhelia, circumzenith arc, and part of the parhelic circle intensity. Singly oriented columns made the tangent arcs, supralateral and infralateral arcs, part of the parhelic circle intensity, and some of the fainter, unlabeled halos. See Appendix G for other, small crystal populations used in the simulation.

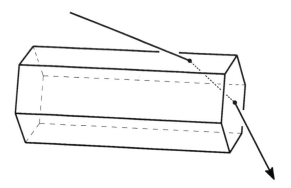

**Figure 2-13.** Ray path for the supralateral arc.

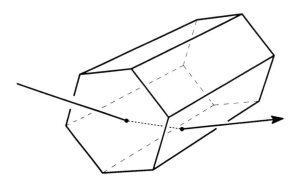

**Figure 2-14.** Ray path for the infralateral arcs.

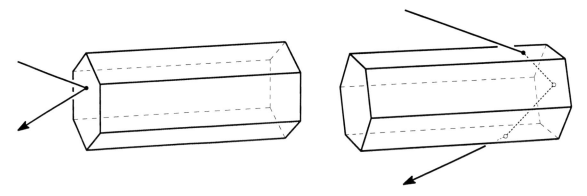

**Figure 2-15.** Two common ray paths for the parhelic circle when caused by singly oriented columns. (Left) External reflection. (Right) Internal reflection. The ray enters a prism face, reflects internally from a basal face, and exits the prism face opposite the entry face. Recall that open circles indicate ray encounters with hidden faces; solid circles, encounters with visible faces.

### Display 2-4
### South Pole, January 17, 1986

**Other halos from singly oriented columns.** As already explained, singly oriented columns cause the tangent arcs, supralateral and infralateral arcs, and some parhelic circle intensity. They can cause several other, rare halos as well, halos found only in the finest displays. Display 2-4, shown on the next several pages, is such a display, with the sky seemingly filled with white and colored arcs.

The crystals collected during the display were beautifully formed plates and columns. The plate crystals caused the parhelia, circumzenith arc, and some of the parhelic circle intensity. The column crystals caused the remaining halos, except perhaps the faint 22° halo. To do so, they must have been almost perfectly horizontal.

How close to horizontal were they? The column crystals used in making Figure 2-18, which successfully simulates the display, had tilts of only 0.15°. If the tilts are increased to 1.0°, the resulting simulation, Figure 2-19, fails noticeably. If instead the tilts are increased to only 0.5°, as in Figure C-20 of Appendix C, the simulation is better but still not quite right. Apparently most of the column crystals in the real display, if they had any tendency at all to be horizontal, were within half a degree of being perfectly horizontal.

The labeled halos in Figure 2-18 are those caused by the singly oriented columns. All of them can be seen in the photographs. The upper tangent arc is dazzling, the supralateral and infralateral arcs are clear, and the parhelic circle is complete and sharp. The other labeled halos are faint in the photographs, and because they are rare, you may wish to skim the following discussion of them. Just remember that should you ever see a display like this, much would be hidden behind the scenes that is no less remarkable than the halos themselves. The crystals would be beautifully formed solid prisms, the singly oriented columns would have almost unbelievably small tilts, and some of the ray paths would be dauntingly complex. The splendor of the halos would be a consequence of the flawless crystal geometry and the precise horizontal orientations.

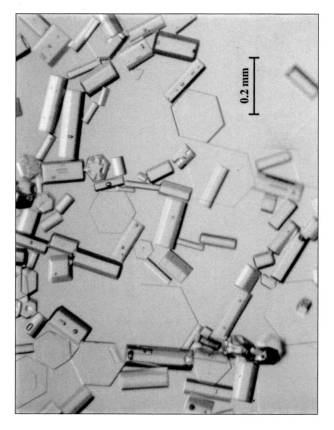

**Figure 2-16.** Some crystals collected during Display 2-4, about five minutes after the halo photographs were taken.

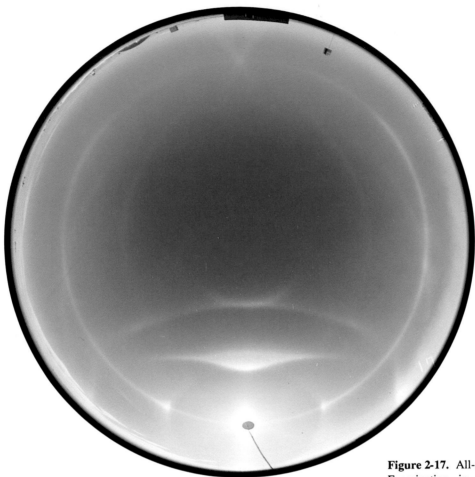

**Figure 2-17.** All-sky view, Display 2-4. Examination in direct sunlight may bring out faint halos in the photograph.

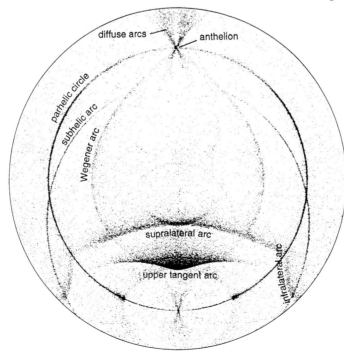

**Figure 2-18.** Simulation of the display. Singly oriented columns made the labeled halos. Other crystal shapes and orientations were also used in making the simulation. For a comparable simulation using singly oriented columns alone, see Figure C-13 of Appendix C.

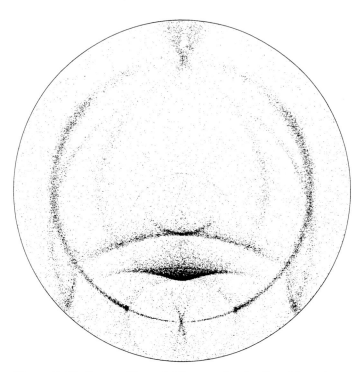

**Figure 2-19.** Same as Figure 2-18 except that the tilts of the column crystals have been increased from 0.15° to 1.0°. The resulting halos are more diffuse than those in the real display.

**Figure 2-20.** Sunward view, Display 2-4. The short streaks were made by light refracted from crystals close to the camera.

**(Upper) Wegener arc.** A ray for the Wegener arc enters a prism face of a singly oriented column and exits an alternate prism face, like the ray for the upper tangent arc, but the Wegener arc ray has an intervening internal reflection from a basal face.

When the crystal axis is nearly perpendicular to the incoming ray, the ray within the crystal just grazes the reflecting face, and the outgoing ray contributes to both the Wegener arc and the upper tangent arc. Thus the Wegener arc coincides with the upper tangent arc directly above the sun.

A lower Wegener arc is also possible (Appendix C). Like the lower tangent arc, it would be below the horizon in Display 2-4.

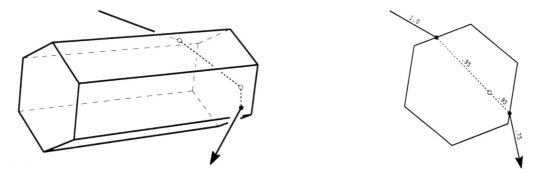

**Figure 2-21.** (Left) Ray path for the (upper) Wegener arc. (Right) End view of the same ray path. Compare with the upper tangent arc ray path, Figure 2-7.

**Figure 2-22.** View to the right of the sun, Display 2-4, with parhelic circle parallel to the horizon, subhelic arc crossing the parhelic circle diagonally from lower center to upper right, and Wegener arc faint in the center and upper right. The bright halos at the left are the upper tangent arc, supralateral arc, right infralateral arc, right parhelion, and circumzenith arc.

**Subhelic arc.** Most subhelic arc rays enter a basal face, reflect internally from a prism face, reflect internally from an alternate prism face, and exit the other basal face. The pair of reflections deviates the ray by 120° as seen from the end of the crystal. The ray paths are therefore much like those for the 120° parhelia, but with different crystal shape and orientation. Yet until recently, only Wegener recognized that the subhelic arc was possible. As a result, some subhelic arcs have probably been mistaken for "Hevel's halo" (Chapter 11).

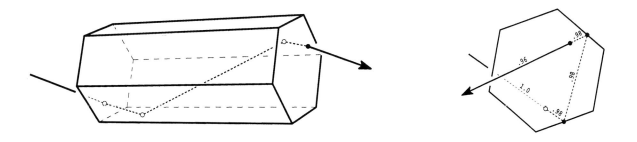

**Figure 2-23.** Common ray path for the subhelic arc.

**Figure 2-24.** View looking away from the sun, Display 2-4. The halos are the parhelic circle, anthelion, diffuse arcs, Tricker arc, subhelic arc, and Wegener arc. Compare Figure C-21.

**Tricker arc.** Although the Tricker arc is normally due to singly oriented columns, it is weak in Figure 2-18 and is not labeled. The simulation in Figure C-21 (Appendix C), which used more crystals, shows it better. The Tricker arc is also faint in the photographs, the best being Figure 2-24.

Two ray paths for the Tricker arc are shown below. Their geometry is much like that of subhelic arc ray paths, except that the Tricker arc paths have an internal reflection from a basal face. Like rays for the subhelic arc and for 120° parhelia, Tricker arc rays are deviated by 120° as seen from the end of the crystal.

If the axis of the upper crystal in the figure were nearly perpendicular to the incoming ray, the ray within the crystal would just graze the basal face, and the outgoing ray would contribute to both the Tricker arc and the subhelic arc. The two halos should therefore have a point in common. In fact, they are tangent at the point, as shown in Figure C-21, but both are weak there.

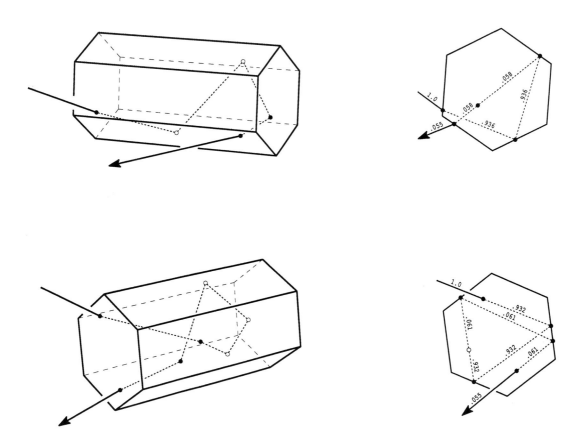

**Figure 2-25.** Two of many ray paths for the Tricker arc.

**Diffuse arcs.** Ray paths for the diffuse arcs are shown below. A halo results from each, but the two halos, referred to as diffuse A and diffuse B, overlap substantially and are not easily distinguished [Greenler and Tränkle, 1984]. Despite their rarity, the diffuse arcs can be rather bright.

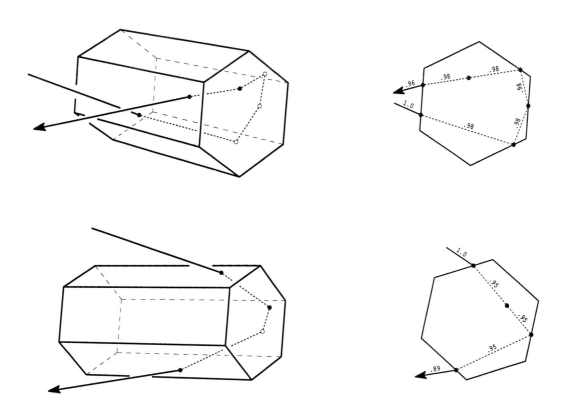

**Figure 2-26.** Ray paths for the diffuse arcs. (Top) Diffuse A. (Bottom) Diffuse B.

**Anthelion.** In the photographs, the anthelion is the white spot on the parhelic circle opposite the sun. (This point on the celestial sphere is called the anthelic point.) An examination of ray paths that make the anthelion in the simulation discloses no new ray paths. That is, the anthelion is just the combined intensity of the other halos that happen to pass through the anthelic point. Of these, the diffuse arcs are the major contributor, as shown in Table 5 of Appendix E.

According to the preceding analysis, the anthelion arises in singly oriented columns. However, the anthelion has been reported in the absence of other column crystal halos [Moon, 1981; Drewry and Rees, 1987].

In the simulations and photographs in this chapter are faint halos not yet mentioned. They are described in Chapter 3.

CHAPTER 3

HALOS FROM PARRY ORIENTED CRYSTALS

Using crystal samples, halo photographs, and simulations, I argued in Chapter 2 that column crystals sometimes maintain their axes horizontal as they fall. In Chapter 3, I will now claim that, in addition, column crystals sometimes maintain two prism faces horizontal. Such crystal orientations are called Parry orientations. Your intuition may doubt the reality of Parry orientations, but simulations will predict characteristic halos from them, and photographs will confirm the existence of the halos. All such halos, however, are rare; the displays in this chapter are extraordinary.

**Display 3-1**
**South Pole, January 21, 1986**

The simulation below shows the halos theoretically expected from column crystals having nearly perfect Parry orientations, that is, from columns having two prism faces almost exactly horizontal. Except for the Parry infralateral arcs, all of the halos labeled in the simulation were present in Display 3-1, shown on the next several pages.

The parhelic circle, subhelic arc, and Tricker arc can form in Parry oriented columns as in this simulation, or in singly oriented columns as described in Chapter 2. Since their ray paths in Parry oriented columns are similar to those described earlier, these three arcs will not be treated again here. The remaining labeled halos are discussed in the following pages.

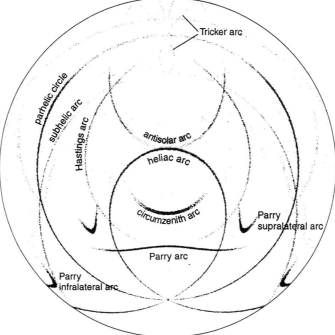

**Figure 3-1.** Simulation showing halos theoretically expected from Parry oriented columns. These abnormally strong halos resulted from using large numbers of crystals. The unlabeled halos have never been reported.

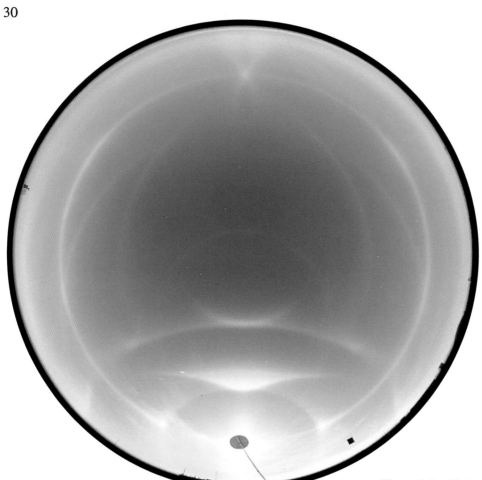

**Figure 3-2.** All-sky view, Display 3-1.

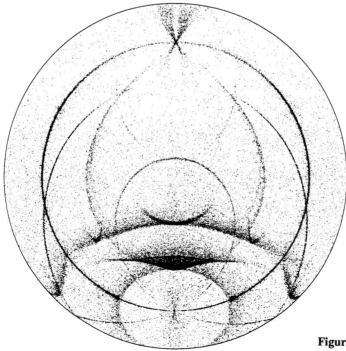

**Figure 3-3.** Simulation of the display.

**Figure 3-4.** Fisheye view of the anthelic region, Display 3-1. The halos are the parhelic circle, 120° parhelia, anthelion, diffuse arcs, Tricker arc, subhelic arc, Wegener arc, heliac arc, antisolar arc, and perhaps faint Hastings arc. See Figure C-21 and Figure C-31.

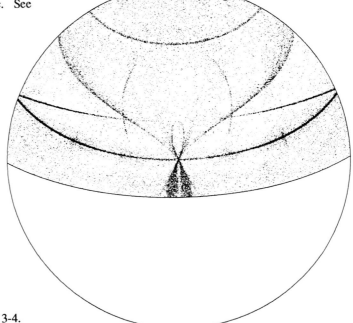

**Figure 3-5.** Simulation of Figure 3-4.

**Figure 3-6.** Another view of Display 3-1. The parhelic circle is parallel to the horizon, and the diffuse arcs are at right. The faint diagonal halos crossing in the upper center are the subhelic arc and Wegener arc.

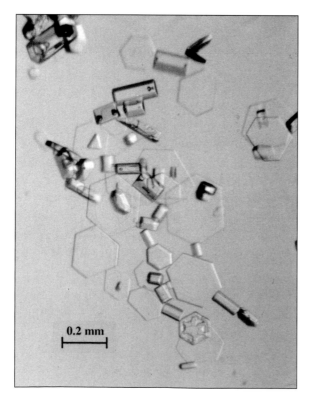
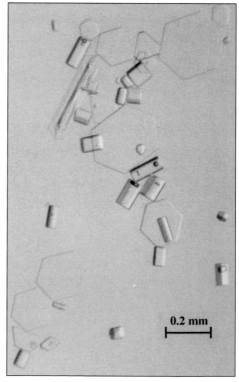

**Figure 3-7.** Some crystals collected during the display.

**(Upper suncave) Parry arc.** When the upper tangent arc is well developed, look for the Parry arc. It often looks like an upper boundary for the upper tangent arc, as in the present display. Display 2-2 also had a Parry arc, but poor; Display 2-3 and Display 2-4 had better ones. The simulations of those three displays also contain Parry arcs. In the simulation in Figure 2-18, for example, four percent of the column crystals were given nearly perfect Parry orientations.

A ray for the Parry arc enters the top prism face of a Parry oriented column and exits an alternate prism face. This ray also contributes to the upper tangent arc; compare Figure 3-8 with Figure 2-7. The Parry arc therefore lies within the upper tangent arc, though not necessarily within the bright part; more on this in Chapter 9.

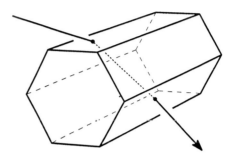

**Figure 3-8.** Ray path for the Parry arc. All of the crystals shown in the ray path figures of Chapter 3 are Parry oriented columns; that is, two prism faces are nearly horizontal.

**Circumzenith arc.** The circumzenith arc can form in Parry oriented columns as well as in oriented plates as described in Chapter 1. In either case the ray enters the top horizontal crystal face and exits a vertical face, but for the circumzenith arc caused by Parry orientations, the entry face is a prism face and the exit face is a basal face. This ray path also contributes to the supralateral arc, so the circumzenith arc lies within the supralateral arc. Compare Figure 3-9 with Figure 2-13.

**Parry supralateral arcs.** A ray for the Parry supralateral arcs enters an upper sloping prism face and exits a basal face.[1] The Parry supralateral arcs therefore lie within the supralateral arc; compare Figure 3-10 with Figure 2-13.

The Parry supralateral arcs are essentially new to the halo literature. Nevertheless, during four austral summers in Antarctica I saw them fairly often, though usually as enhancements of the supralateral arc rather than as the well defined curves in Figure 3-2.

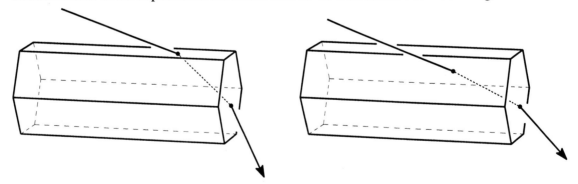

**Figure 3-9.** Ray path for the circumzenith arc when formed by Parry oriented crystals.

**Figure 3-10.** Ray path for the Parry supralateral arcs.

---

1 Though less likely, the ray path that enters a lower sloping prism face and exits a basal face is also possible. The resulting arcs are faintly visible outside the Parry infralateral arcs in Figure 9-2, but they have never been seen in reality. They lie within the supralateral arc, though certainly not within the bright part.

**Parry infralateral arcs.** A ray for the Parry infralateral arcs enters a basal face and exits a lower sloping prism face. These arcs therefore lie within the infralateral arcs. Compare Figure 3-11 with Figure 2-14.

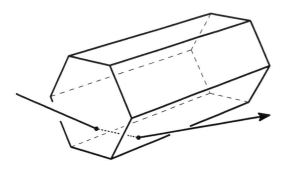

**Figure 3-11.** Ray path for the Parry infralateral arcs.

**Heliac arc.** The simplest rays for the heliac arc reflect externally from a sloping prism face. More complicated paths are also possible.

In Antarctica I saw the heliac arc frequently. Indeed, it was often more conspicuous than the better known Parry arc. In addition to the heliac arc of the present display, faint heliac arcs appear in Display 2-3 and Display 2-4, where they form giant X's at the sun. (When I said earlier that the halos of this chapter were rare, I was referring to my experience in temperate climates.)

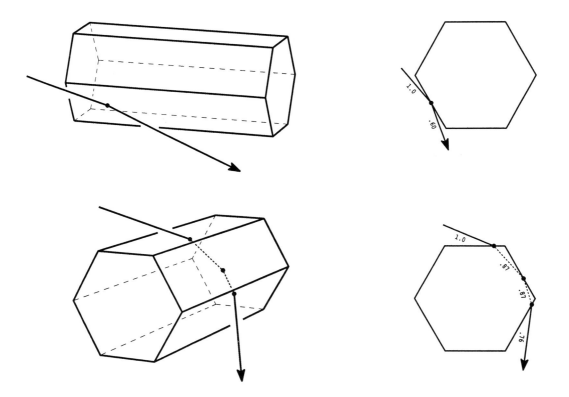

**Figure 3-12.** (Left) Two common ray paths for the heliac arc. (Right) End views of the same ray paths.

**Antisolar arc.** Ray paths for the antisolar arc are like those for the heliac arc but with an internal reflection from a basal face. The antisolar arc is therefore tangent to the heliac arc, much as the Tricker arc is tangent to the subhelic arc. Long complicated ray paths are common for the antisolar arc.

The antisolar arc is clear in the original slides of Figure 3-2 and Figure 3-4, but it may be too faint to survive reproduction.

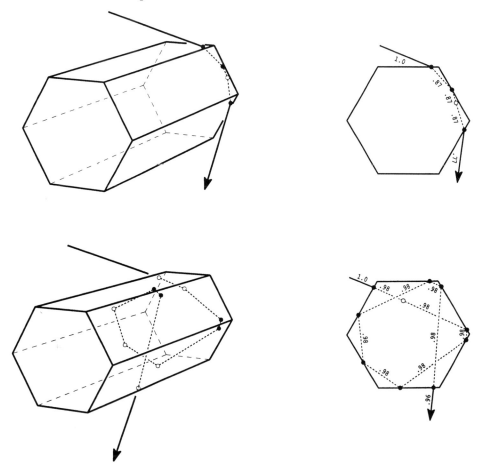

**Figure 3-13.** Two common ray paths for the antisolar arc. Compare the upper diagram here with the lower diagram in Figure 3-12.

**Hastings arc.** A ray for the Hastings arc enters the top prism face and exits an alternate prism face, like the ray for the Parry arc, but the Hastings arc ray has an intervening internal reflection from a basal face.[2] The Hastings arc is tangent to the Parry arc and lies within the Wegener arc. The Hastings arc is nearly indistinguishable from the Wegener arc, but the original slide of Figure 3-4 does seem to show a distinct Hastings arc at the upper part of the Wegener arc.

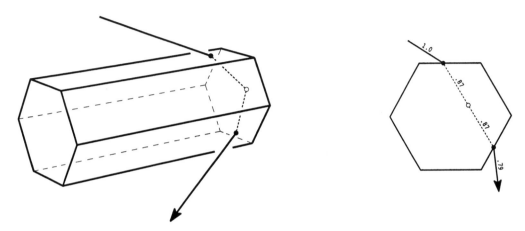

**Figure 3-14.** Ray path for the Hastings arc. Compare with Figure 3-8 and Figure 2-21.

Most of the halos mentioned in the first four chapters appear in Display 3-1. The simulations in Figure 3-3 and Figure 3-5 used singly oriented columns, Parry oriented columns, oriented plates, and randomly oriented crystals to reproduce the many halos. The column orientations, as you may have guessed, had to be nearly perfect (see Appendix G). As far as I know, this display is the finest ever photographed. I await your counterexamples!

---

2 Eventually we will recognize four different Parry arcs (Displays 3-1, 3-4, 6-7, and 6-8). Each in principle has its own Hastings arc (Appendix C), whose ray path is like that for the corresponding Parry arc but with an internal reflection from a basal face. Thus the Hastings arc mentioned above is technically the Hastings arc of the upper suncave Parry arc. The other three Hastings arcs have never been seen in reality.

**Display 3-2**
**South Pole, January 13, 1986**
**Display 3-3**
**South Pole, January 17, 1986**

Scanning the normally featureless South Pole horizon, we detect an odd low cloud, almost like dust kicked up by a distant herd of animals. As the cloud approaches, colored or white patches in it — bits of halos — show that it contains ice crystals. Suddenly the crystal swarm envelopes us and creates a breathtaking array of halos. Within ten minutes the swarm passes, leaving only memories of the halos.

Such is a common scenario for elaborate South Pole displays. Even there, great displays are unusual, but a handful do occur each summer. Mid-January 1986 saw half a dozen displays only slightly inferior to Display 3-1. Among them were the two displays of this section.

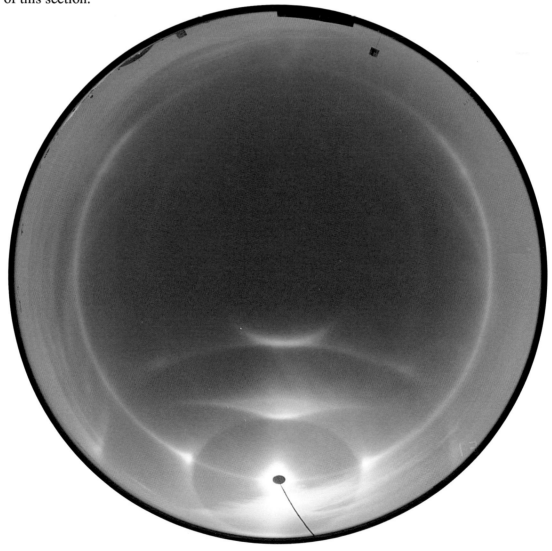

**Figure 3-15.** All-sky view, Display 3-2. Crystals from the display are shown at the upper right in Figure 1-3.

**Figure 3-16.** Bright upper tangent arc and circumzenith arc, Parry arc, parhelic circle, subhelic arc, Wegener arc, heliac arc, supralateral arc, poor Parry supralateral arcs, infralateral arc at lower left, and 22° halo at bottom. (Display 3-3, about eight hours after Display 2-4)

**Figure 3-17.** Some crystals collected during the display above.

**Associating the crystals with the halos.** You may have questioned whether ice crystals collected during a halo display can be reliably associated with the halos. The crystals simply fall into a Petri dish left in the open; what if the halos are forming in crystals higher in the atmosphere, out of reach? To rule out this possibility, the halos must pass a crucial test: They must show up against the snow surface (Figure 7-9) or in front of a nearby structure (Figure 1-1, Figure 2-3). At the South Pole a 25-foot tower was erected as a background for detecting halos, especially those overhead. Halos that appear in front of the tower or other nearby structures are obviously forming in crystals close at hand.

When halos in a display are strong enough to appear in front of a structure, and when the crystal swarm is heavy, I am confident that the halos are being caused by crystals like those being collected. The weaker halos in the display cannot be expected to show up in front of the structure, but if they come and go along with the strong halos, they are probably forming in the same crystals. In the best Antarctic displays, where the crystals and halos appeared simultaneously, there was usually little doubt that the halos arose in crystals right around us. In weak displays, I sometimes could not confirm that the halos were nearby, and crystal samples from such displays do not appear in this book.

**Figure 3-18.** Tower used as a background to detect nearby halos. No halos are present.

### Display 3-4
### South Pole, February 16, 1986

The Parry arc described previously is technically the (upper) suncave Parry arc. Its ray path was shown in Figure 3-8, where the sun elevation was 20°. If the sun is lower, the ray path shown below becomes possible and produces the (upper) **sunvex Parry arc**, the faint V-shaped arc in the photograph and simulation on the facing page. A heliac arc is also visible, consistent with Parry orientations.

The crystal sample for the display consisted of long columns. Similarly shaped crystals were used in making the simulation, some singly oriented and some Parry oriented. The simulation fails to reproduce the conspicuous gap in the heliac arc but seems otherwise adequate. Evidently the real crystals had orientations close to those in the simulation.

**Homogeneous crystal samples.** Because the crystal sample contained only long columns — no other kinds of crystals — and because the halos formed close at hand, the halos must have been caused by long columns. Homogeneous crystal samples, such as this one, yield rather definitive results: they reveal the crystal type responsible for the halos, and then simulations can discover the orientations. Unfortunately, homogeneous samples are uncommon in good halo displays. Display 1-1, Display 1-3, and Display 2-1, as well as the present display, are special because of their homogeneous samples.

**Effect of crystal shape on halo intensities.** The longer and thinner are the column crystals in a display, the weaker will be the halos whose ray paths involve basal faces, relative to the halos whose ray paths require only prism faces [Pattloch and Tränkle, 1984]. In the present display, for example, the crystals are long and thin. In both the simulation and the photograph, the upper tangent arc is bright but the supralateral arc, whose ray path involves a basal face, is faint. Similarly, the Parry arc is visible but the Parry supralateral arcs are absent. Compare Display 3-1, where the column crystals are stubbier and where the halos requiring basal faces are clear.

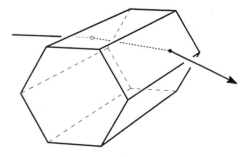

Figure 3-19. Ray path for the sunvex Parry arc. The ray enters one upper sloping prism face and exits the other.

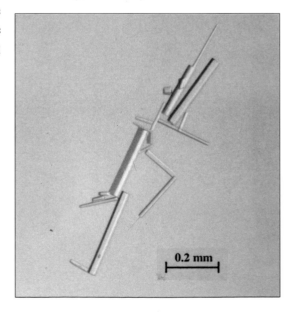

0.2 mm

Figure 3-20. Some crystals collected during the display.

**Figure 3-21.** Upper tangent arc, sunvex Parry arc, supralateral arc, and heliac arc. The sunvex Parry arc is the faint V-shaped arc above the bright part of the upper tangent arc. (Display 3-4)

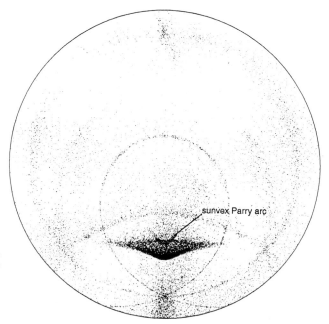

**Figure 3-22.** Simulation of the display. Singly oriented columns made the upper tangent arc and the faint supralateral arc. Parry oriented columns made the Parry arcs and heliac arc.

## Display 3-5
## Vostok Station, Antarctica, January 19, 1991

**How do Parry orientations occur?** Parry orientations have long been thought to be the cause of Parry arcs, but people have wondered how such orientations could occur. That is, how could a falling crystal maintain two prism faces horizontal? Might it, for example, occur in a cluster? Or might it be tabular shaped, with two prism faces exaggerated?

The displays in the present chapter, as well as Display 2-3 and Display 2-4 in the preceding chapter, contain many examples of halos arising in Parry oriented crystals. None of the crystal samples for these displays contained clusters. And although a column crystal occasionally had two opposite prism faces somewhat larger than the other four (e.g., Figure 2-16), I would not call these crystals tabular — certainly they bore little resemblance to the figure below. The absence of clusters and tabular crystals in the crystal samples shows that ordinary columns by themselves can assume Parry orientations — no clusters or exotic shapes are needed. Parry orientations seem to be a normal falling mode for ordinary columns.[3] Perhaps a certain fraction of column crystals naturally fall with Parry orientations.

Now Display 3-5 adds an intriguing qualification. The display is dominated by the parhelia, but the remarkable feature is the presence of halos from Parry orientations — the sunvex Parry arc and heliac arc, together with suggestions of the suncave Parry arc, Parry supralateral arcs, and antisolar arc — in the near absence of halos from singly oriented columns. Apparently nearly all of the horizontal column crystals had Parry orientations. Certain crystal shapes and sizes, or perhaps atmospheric conditions, evidently favored Parry orientations. Unfortunately, except for the parhelia and probably the circumzenith arc, the halos were not at low level, and no meaningful crystal sample could be obtained. The requirements for Parry orientations therefore remain unclear.

**Figure 3-23.** Cluster of column crystals proposed to account for Parry orientations. No such clusters were seen during Parry arc displays.

**Figure 3-24.** Tabular crystal proposed to account for Parry orientations. Such crystals seem to be very rare. (adapted from Tricker [1970])

3 Since crystal samples from most Parry arc displays contain plates as well as columns, one might ask whether plates, rather than columns, are the Parry oriented crystals. But the evidence from the first two chapters suggests that only columns can maintain their axes horizontal. Also, simulations made with Parry oriented plates predict somewhat unrealistic displays. Finally, Display 3-4 shows that columns can indeed assume Parry orientations. Still, more crystal samples from displays like Display 3-4 would be welcome.

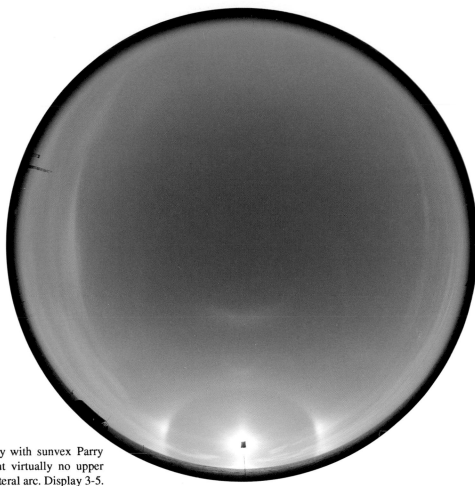

**Figure 3-25.** Display with sunvex Parry arc and heliac arc but virtually no upper tangent arc or supralateral arc. Display 3-5. (Photograph by G. P. Können)

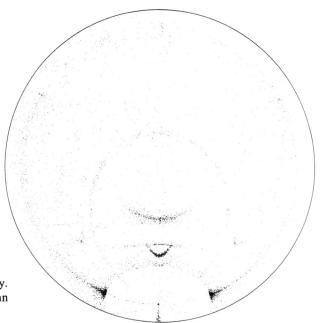

**Figure 3-26.** Simulation of the display. Almost 90% of the horizontal column crystals had Parry orientations.

**Figure 4-1.** The 22° and 46° halos, parhelia, circumzenith arc, and poor upper tangent arc. The two faint thin circles inside and just outside the 22° halo are not real; they result from aiming the wide angle lens at the unblocked sun. Eau Claire, Wisconsin, January 29, 1977.

# CHAPTER 4

## THE 22° AND 46° HALOS

Appearing every few days or so, the circular halo of angular radius 22° is familiar to skywatchers. The "ring around the moon," for example, is just a 22° halo formed in moonlight. The 46° halo, a much larger circular halo, is seen far less often.

Figure 4-2 is a simulation showing the halos theoretically expected from randomly oriented column crystals. Unless you have experimented with a glass prism as suggested in Chapter 1, the simulation may come as a surprise. Instead of a rather uniformly lit sky, the simulation predicts two circular halos — the 22° and 46° halos.[1]

As in the preceding chapters, the computer can be asked to find ray paths responsible for lighting specified regions of sky in the simulation. A light ray for the 22° halo turns out to be like parhelion and tangent arc rays; it enters a prism face of a crystal and exits an alternate prism face. A ray for the 46° halo is like circumzenith arc, supralateral arc, and infralateral arc rays; it enters a prism face and exits a basal face, or vice versa.

While in fact the ray paths for the circular halos have long been known, the responsible crystals and orientations are less certain. Figure 4-2 suggests that randomly oriented crystals are the cause of the circular halos, and by and large this has been a standard assumption. A second assumption has been that the crystals that will orient randomly are equidimensional, or perhaps very small or very large. (An equidimensional crystal is intermediate in shape between a plate and a column.) Both assumptions require qualification, and the rest of this chapter is devoted to examining them in the light of simulations as well as real displays. If these issues do not concern you, you should skip to the next chapter.

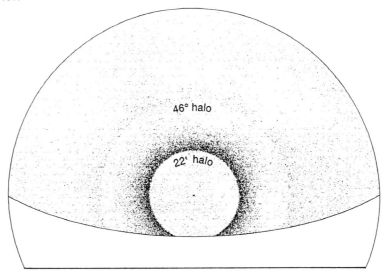

**Figure 4-2.** Simulation showing halos theoretically expected from randomly oriented column crystals. The sun elevation is 20°.

---

1 A glass prism provides some insight for the 22° halo, but for the 46° halo the large refractive index of glass prohibits the ray path.

**Circular halos do not require random crystal orientations.** Figure 4-2 was made with randomly oriented columns; they made 22° and 46° halos. In Figure 4-3 the same crystals were given a slight tendency to be horizontal — their tilts were 30°.[2] These poorly oriented columns again made two circular halos. The inner halo, though stronger at the top and bottom, is still a 22° halo rather than upper and lower tangent arcs. And the outer halo is indistinguishable from the 46° halo made by the randomly oriented crystals in Figure 4-2. Perfectly random orientations are not necessary for the circular halos, especially for the 46° halo.[3]

In Figure 4-4 the crystal tilts were reduced further, to 5°. These better oriented columns transformed the 22° halo of Figure 4-3 into an upper tangent arc, but they had less effect on the outer halo, which is still nearly circular, showing just a suggestion of supralateral and infralateral arcs. To give easily identifiable supralateral and infralateral arcs, the crystal tilts must be reduced to about 2°, as in Figure 4-5.

The simulations show that the 46° halo can arise in crystals with unexpectedly small tilts, but they also show that as the tilts decrease, the 46° halo does eventually evolve into something else, namely, the supralateral and infralateral arcs. The evolution is subtle, and no sharp line separates the 46° halo from the supralateral and infralateral arcs. In practice, the distinction is often missed entirely. The outer halo in Figure 4-4, for example, might easily be misidentified as a 46° halo, but its not quite circular shape shows that it consists of poorly defined supralateral and infralateral arcs. The accompanying halos, since they give information about crystal orientations, can also help in the identification. The strong upper tangent arc, and especially the absent 22° halo, are signs that the halo in question consists of supralateral and infralateral arcs. Conversely, a strong 22° halo and a weak or absent upper tangent arc would confirm the identification as a 46° halo.

---

2 How far are these crystals from being randomly oriented? As explained in step 3 of Appendix F, 68% of them are within 30° of horizontal, and 95% within 60°, whereas for randomly oriented crystals, 50% are within 30° of horizontal, and 87% within 60°.

3 The necessity of random crystal orientations for the circular halos was questioned by Fraser [1979]. For the 46° halo, the same question had been raised earlier by Hastings [1920].

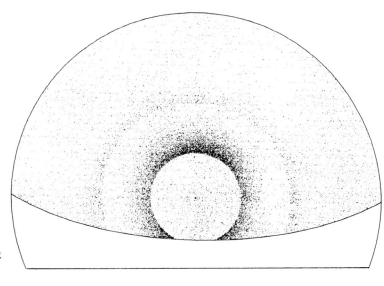

**Figure 4-3.** Same as Figure 4-2 except that the crystal tilts are 30°.

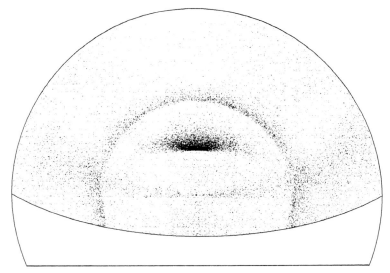

**Figure 4-4.** Crystal tilts of 5°.

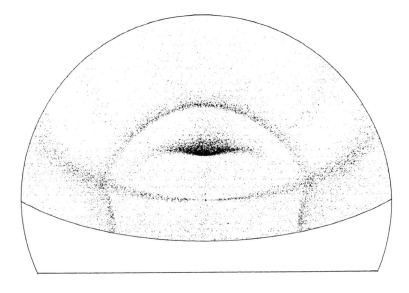

**Figure 4-5.** Crystal tilts of 2°.

### Display 4-1
### South Pole, January 1, 1986

**Circular halos do not require equidimensional crystals.** This display had a lower tangent arc on the horizon, a poor upper tangent arc, and hints of halos in the 46° region. The predominant halo, however, was the 22° halo.

The crystal sample consisted of large columns and clusters of columns. The 22° halo therefore could not have formed in equidimensional crystals; there were none. Instead, it must have formed in crystals like those in the sample. These columns and clusters evidently had only a slight tendency to be horizontal, as shown by the strength of the 22° halo and the poorly quality of the upper tangent arc.

This display and its crystals are typical of summertime South Pole displays when the crystals originate at high level. The halos are mediocre, often only the 22° halo being at all conspicuous. The crystals are large poorly formed columns or column clusters.

The usual 22° halo display in temperate climates is not much different. An unimpressive 22° halo, perhaps enhanced at the top and bottom, may be the only halo present. Like Display 4-1 at the South Pole, these displays, I believe, are caused by large imperfect columns or column clusters.

Liljequist [1956] reached the same conclusion from halo and ice crystal observations at Maudheim on the west Antarctic coast in 1950-1951. Even earlier, after sampling crystals in high clouds over Germany, Weickmann [1948, p. 65] wrote, "On our flights we learned to know the 22° halo as the typical halo of cluster and hollow crystals..."

**Figure 4-6.** The 22° halo. (Display 4-1)

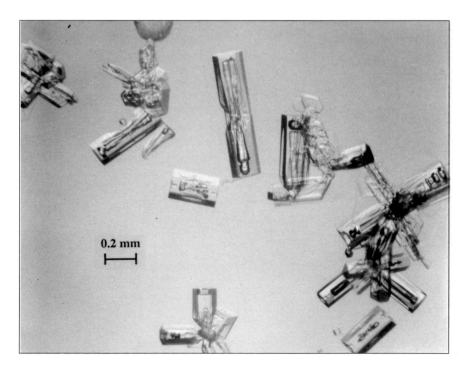

0.2 mm

**Figure 4-7.** Some crystals collected during the display.

## Display 4-2
### Fairbanks, Alaska, February 24, 1984

**Equidimensional crystals need not result in circular halos.** Many of the crystals collected during this display were well-formed equidimensional prisms. They look like good halo makers, and had they oriented randomly, I would have expected a 22° halo. A few sparkles did appear where the 22° halo should have been, but not enough for the 22° halo to show in the photograph. Instead, the dominant halos were the parhelia. The equidimensional crystals apparently fell like oriented plates, with their basal faces more or less horizontal, and made parhelia.

If this were an isolated observation, I would rationalize it in various ways, but other halo and crystal observations agree: rather than orienting randomly, equidimensional crystals — and perhaps even short columns — can fall like oriented plates. (This may be easier to accept if you recall Parry orientations; if column crystals can sometimes maintain two prism faces horizontal, perhaps equidimensional crystals can sometimes maintain their basal faces horizontal.) If the parhelia and circumzenith arc are to be attributed to oriented plates, as in Chapter 1, then equidimensional crystals must sometimes be regarded as "plates".

Well, what crystals *do* make the 22° halo? We have seen that large imperfect columns and column clusters are responsible for the 22° halos in many run-of-the-mill displays. But what about the 22° halos in good displays? I do not know, in spite of having looked at crystals during hundreds of halo displays. Several displays in the preceding chapters were selected in part because there was no 22° halo; its absence simplified the task of relating crystal types with halo types. In spite of its symmetry and familiarity, the 22° halo is not so simple. Had I understood it better, I would have begun the book with it.

**Figure 4-8.** Parhelia. (Display 4-2)

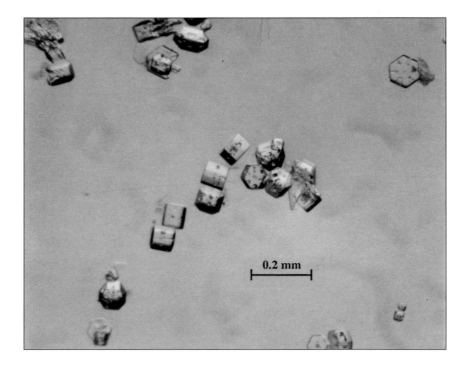

**Figure 4-9.** Some crystals collected during the display.

CHAPTER 5

## WHY ARE THE RARE HALOS RARE?

If you have never watched the sky, you will be surprised to find that halo displays occur frequently — more than 100 days a year in my experience. I am referring not to the Antarctic interior, where conditions are exceptional, but rather to my experience in Wisconsin, which is probably representative of temperate conditions. There you will often see 22° halos, parhelia, circumzenith arcs, sun pillars (Chapter 6), and poorly defined tangent arcs. You must be more patient, however, to see 46° halos, parhelic circles, 120° parhelia, supralateral arcs, infralateral arcs, and good tangent arcs. And a lifetime of skywatching may not capture a display as fine as, say, Display 2-4; Wegener arcs, subhelic arcs, diffuse arcs, heliac arcs, etc., just do not seem to occur under normal atmospheric conditions.

Why are the rare halos rare? Consider the simulation shown in Figure 5-1, and rephrase the question. Why are there so few displays as good as the simulation? This chapter suggests some answers:

**Crystal numbers.** In the simulation some halos are weaker than others.[1] Naturally the weak ones will be seen less often, since they require large numbers of crystals.

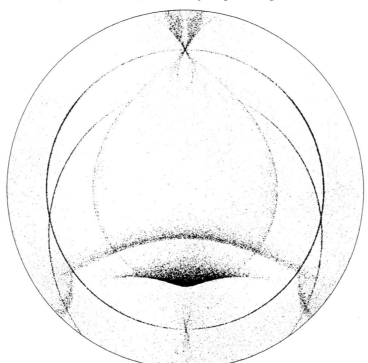

**Figure 5-1.** Simulation showing halos theoretically expected from column crystals having tilts of 0.1°. The sun elevation is 20°.

---

1 Some reasons for weak halos in a simulation:
The ray path has low intensity. Example: Tricker arc ray paths.
Relatively few rays travel the ray path. Example: parhelion ray paths through thin plates.
The outgoing ray is sensitive to small changes in crystal orientation. Example: rays for parhelion tails.

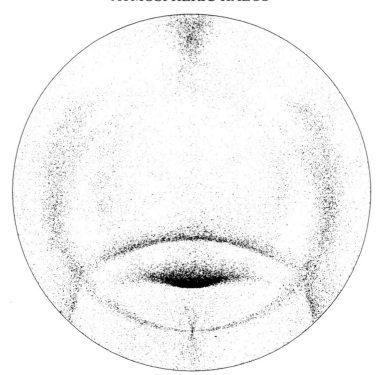

**Figure 5-2.** Same as Figure 5-1 except the crystal tilts are now 3°.
The very rare halos have disappeared.

**Crystal orientations.** Figure 5-1 was made using almost perfectly horizontal columns. Figure 5-2 is the same except that the crystal tilts have been increased to 3°. The very rare halos are gone; only the upper tangent arc, supralateral arc, infralateral arcs, and a bit of parhelic circle near the sun remain. Clearly, the best halo displays require that the singly oriented columns have extremely small tilts.

Such small tilts, however, are not as difficult to attain as intuition may suggest. They sometimes occurred at the South Pole even in a light wind, if the observed halos are indeed an indication.

**Crystal size.** As explained in Appendix F, the simulations do not predict the effect of crystal size on halos. The following crude size estimates are derived largely from crystal samples collected during halo displays. The estimates are imprecise because crystal sizes vary within each sample.

Very small crystals, roughly 0.01 mm, do not make perceptible halos. The ray optics used in halo theory is incorrect at such small scales.

Crystals larger than about 0.04 mm can produce halos, but below about 0.1 mm the crystals do not attain the very small tilts necessary for rare halos (Figures 5-3 and 5-4).

To attain these small tilts, crystals must exceed about 0.1 mm. Yet column crystals that are too long — longer than about a millimeter — are usually poorly formed and slightly asymmetric, and therefore still unable to attain the desired tilts. Some writers think that such large crystals would spin or flutter even if they were well-formed symmetric prisms [Tricker, 1970; Hallett, 1987]. So the crystals that make good halos seem to be confined to a somewhat narrow size range.

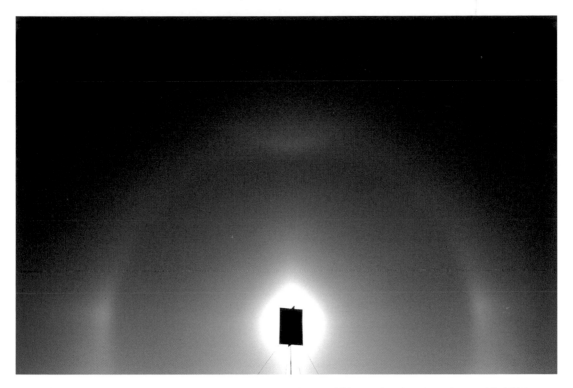

**Figure 5-3.** Unspectacular upper tangent arc, parhelia, and 22° halo. South Pole, February 10, 1986.

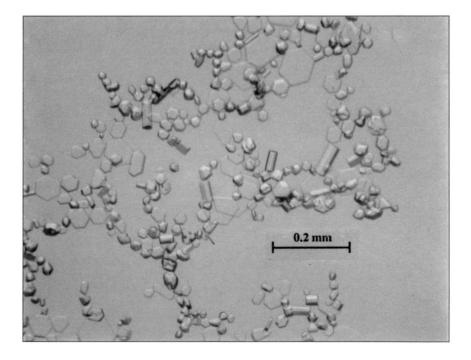

0.2 mm

**Figure 5-4.** Some crystals collected during the display above. These crystals are probably too small to assume the very small tilts needed for sharply defined halos.

**Crystal development.** The ice crystals shown in this book were not randomly chosen. They, after all, are generally the crystals that made good halos. The best ones have virtually perfect planar faces and are very nearly the ideal solid hexagonal prisms that halo theorists postulate. Figure 2-16 is a good example.

The crystal shown below — a "hollow column" — is probably a more common type. Similar columns have been collected by aircraft in high clouds. What halos could such crystals produce if singly oriented?

The large, roughly conical cavities in the ends virtually obliterate the basal faces. These cavities would preclude all halos whose ray paths involve basal faces. Of the many halos in Figure 5-1, only the upper tangent arc could occur. And it would be mediocre, since the basal face cavities and the imperfect prism faces would interfere with the necessary ray paths, and since the slight asymmetry and large size of the crystal might promote excessive tilts. So these crystals would make at best an unimpressive upper tangent arc. If randomly oriented, they would make an equally unimpressive 22° halo.

Indeed, mediocre upper tangent arcs and 22° halos are common, whereas other halos due to column crystals are infrequent.[2] It seems that, at least in clouds that are thin and visible from the ground, atmospheric conditions are usually not conducive to growing and preserving column crystals with complete planar faces, especially basal faces. Poor crystal development — scarcity of well-formed prisms — results in few and mediocre halos, or even no halos.

There are many types of ice crystals in addition to the solid prisms that make the best halos. There are the hollow columns discussed above, there are the familiar dendritic "snowflakes," there are needle crystals, there are complicated spatial crystals, and more. Each crystal type, including the solid prism, is thought to require characteristic conditions of temperature and humidity for its formation (Appendix D). If the conditions are wrong, solid prisms will not form, and good halos will not occur. A cloud of ice crystals is no guarantee of halos.

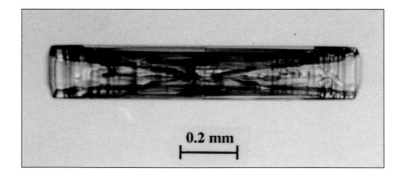

0.2 mm

**Figure 5-5.** Column crystal with a large cavity in each end. Such crystals are common but do not make good halo displays. Vostok Station, Antarctica, January 22, 1991.

---

2 I am referring to sun elevations of about 20°. Poor lower tangent arcs are common when the sun is higher, and sun pillars are common when lower.

**Visibility.** When well-formed crystals do occur, in temperate climates they are apt to be in high cloud. In order to produce halos visible from the ground, the cloud must not be too thick and must not be obscured by lower cloud.

Even when the halo-producing cloud is not actually obscured, atmospheric water droplets, small particles, and poorly formed or small ice crystals can decrease halo visibility. By scattering sunlight in a relatively random fashion, they create a light grey background that camouflages the weak halos, especially the many rare white halos. When located between the sun and the halo-producing ice crystals, they weaken the available sunlight, and when located between the crystals and the observer, they scatter the halo light and further degrade the halos. In general, atmospheric particles that do not make halos, interfere with halos.

For comparison, consider what happens when large, well-formed ice crystals predominate. Much of the sunlight impinging on such a crystal emerges undeviated (it enters one face and directly exits the parallel face), and most of the remainder emerges in only a few directions — the directions of the halos. Such crystals are nearly invisible unless they are in a position to send halo light to the observer. The result is bright halos in a clear sky. Display 6-5 is an example.

**Conclusion.** I believe poor crystal development is the main reason that good halo displays are uncommon. And what separates the best displays from the merely good ones are the extremely small tilts of the column crystals.

Neither the atmospheric conditions necessary for good crystal development nor the conditions for small tilts are fully understood. In particular, I am not sure what it is about Antarctic conditions that makes extraordinary halos.

CHAPTER 6

THE ROLE OF SUN ELEVATION

The shapes of most halos change dramatically with sun elevation. Some halos even disappear when the sun is high, while others are possible only then. The impact of sun elevation adds to the enjoyment (and confusion) of halo watching, since familiar winter halos may be unrecognizable at midday in summer. The simulations in Appendix C show the theoretical dependence of halos on sun elevation, and the present chapter illustrates the dependence using real displays. Only the first two displays were caused by low-level crystals, and only the first is accompanied by a crystal sample.

**Display 6-1**
**Fairbanks, Alaska, October 29, 1992**

**Sun pillar.** A vertical shaft of light sometimes emanates from the rising or setting sun. Known as a sun pillar, it arises in reflections from predominantly horizontal crystal faces, the crystals being oriented plates, singly oriented columns, or Parry oriented columns. Because the crystals need not be well formed to be effective, pillars are common and may be unaccompanied by other halos.

**Figure 6-1.** Pillar below the sun, which is just out of the photograph at the top. No other halos were present. The sun elevation was about 6°. (Display 6-1)

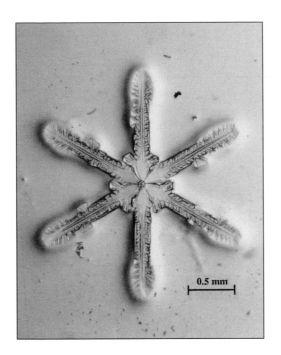

**Figure 6-2.** Typical crystals collected during Display 6-1. Crystals of many sizes and types can make sun pillars, those in this display being large and ornate. The lack of sizable prism faces here is consistent with the absence of other halos in the display. Shown here are not the crystals themselves, but rather plastic replicas, made by allowing the crystals to fall on a glass slide and then lightly coating them with acrylic spray. The top right photograph is a detail of that at the top left.

## Display 6-2
## Barrow, Alaska, May 8, 1979

**Lower tangent arc.** The lower tangent arc was introduced in Display 2-1, but the display was weak and the arc was below the horizon. The few sparkles in Figure 2-3 gave no inkling of how impressive this halo can be. In Figure 6-3, however, the sun is nearly 10° higher, and the lower tangent arc is above the horizon. In brightness it nearly rivals the sun.

Note also the fine 46° halo, clearly distinct from the supralateral and infralateral arcs. The uniform 22° halo and the sharply defined tangent arcs indicate that the 46° halo here was caused by nearly randomly oriented crystals; the halo is not just a combination of poorly defined supralateral and infralateral arcs.

**Figure 6-3.** Bright lower tangent arc at lower left, upper tangent arc and Parry arc, 22° and 46° halos, parhelic circle, faint parhelion, infralateral arc at lower right, faint supralateral arc at upper right, and faint Wegener arc. The asymmetry of the upper tangent arc is due to the wide angle lens and is not real. Temperature +23° F. Display 6-2. (Photograph by Takeshi Ohtake)

## Display 6-3
## Northern New Mexico, February 25, 1988

In this display the sun elevation was about 34°, at least 10° more than in earlier chapters. As a result, the upper tangent arc is flatter, the parhelia have moved out from the 22° halo, and the parhelic circle has shrunk, as it must, in order to pass through the sun and remain parallel to the horizon. Bright 120° parhelia and a fine Parry arc highlight the display.

The special conditions necessary for elaborate halo displays are evidently not confined to the polar regions. But whatever the conditions are, they are rare elsewhere. This display drew crowds outdoors from Los Alamos to Albuquerque.

**Figure 6-4.** Front page of the Santa Fe *New Mexican* on the day after the display.

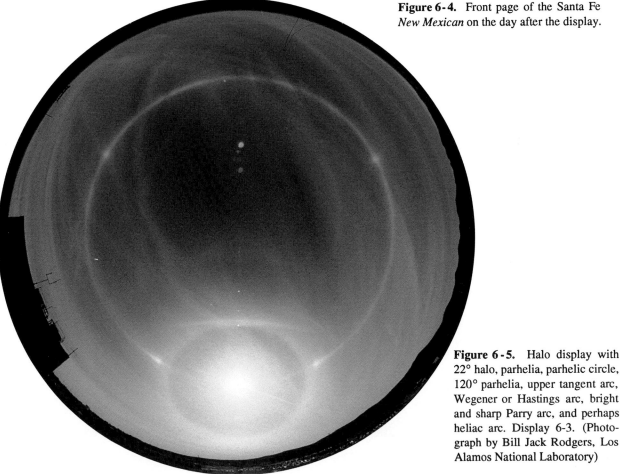

**Figure 6-5.** Halo display with 22° halo, parhelia, parhelic circle, 120° parhelia, upper tangent arc, Wegener or Hastings arc, bright and sharp Parry arc, and perhaps heliac arc. Display 6-3. (Photograph by Bill Jack Rodgers, Los Alamos National Laboratory)

**Display 6-4**
**Atacama Desert, Northern Chile, February 17, 1990**

**Circumscribed halo.** As the sun rises in the sky, the upper and lower tangent arcs bend toward each other at their extremities, and at a sun elevation of 29° the two halos merge to form the circumscribed halo. As the sun climbs further, the circumscribed halo continues to change shape, becoming less distorted and more nearly circular, as shown in Appendix C. In the display below, the sun elevation was about 60°.

Of course, the circumscribed halo arises in the same crystals that make the tangent arcs at lower sun elevations, namely, singly oriented columns. To make the fine halo here, the prism faces must have been well formed and the tilts small. So the absence of other column crystal halos, especially the parhelic circle, is striking. The basal faces of the crystals were evidently ineffective. Perhaps they were poorly formed, or the columns were very long and thin.

The value 29° mentioned above is theoretical. For sun elevations less than about 35°, the sides of the circumscribed halo are normally imperceptible, and separate upper and lower tangent arcs are seen, as in Display 6-3.

**Figure 6-6.** Splendid circumscribed halo; Display 6-4. (Copyright Gosewijn W. J. van Nieuwenhuizen.)

## Display 6-5
### Fairbanks, Alaska, May 8, 1983

The photograph below shows the 120° parhelia and the parhelic circle. The sun was high enough, about 39°, to make the parhelic circle uniform and complete. (Only part is shown here.) By contrast, parhelic circles at low sun typically are weak opposite the sun.

Figure 1-17 helps to explain why. There, at the internal reflection in the lower crystal, intensity is lost to the (unshown) exit ray, a circumzenith arc ray; the circumzenith arc is robbing the parhelic circle in the anthelic region. For sun elevations greater than about 32°, however, the internal reflection becomes total, the circumzenith arc disappears, and the parhelic circle intensity is restored. This argument applies to parhelic circles caused by oriented plates. For low-sun parhelic circles caused by singly oriented columns, anthelic intensity is lost to the supralateral arc rather than to the circumzenith arc.

No cloud is visible in the photograph; only the halos betray the presence of ice crystals. The simulation, except that it fails to explain the blue color of the sky, is an excellent match for the photograph. Apparently the assumptions of the halo theory used in the simulations — simple hexagonal prisms, ray optics, single scattering — are very nearly correct here. In the usual halo display, however, where clouds are evident, the theory incompletely describes the light from the sky.

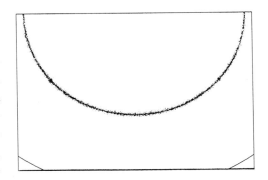

**Figure 6-7.** Simulation of the display. Oriented plates made the halos.

**Figure 6-8.** The 120° parhelia and parhelic circle opposite the sun. (Display 6-5)

**Figure 6-9.** Halo display at very high sun. (Display 6-6)

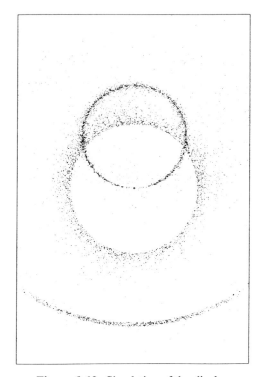

**Figure 6-10.** Simulation of the display. Oriented plates and randomly oriented crystals made the halos.

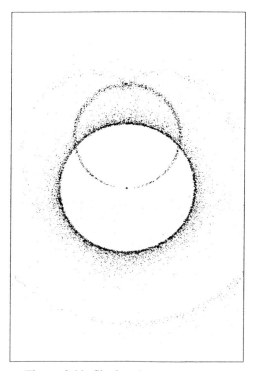

**Figure 6-11.** Singly oriented columns made the halos.

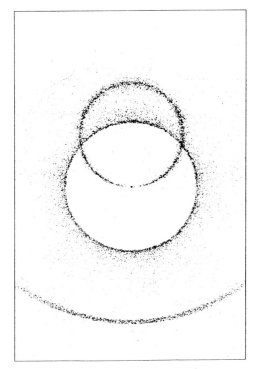

**Figure 6-12.** Oriented plates and singly oriented columns made the halos.

## Display 6-6
## Tooele, Utah, June 16, 1980

In this display the sun was very high, about 72°. There were three halos: Low in the sky was a colored arc, weak in the photograph but strengthening later as the ice crystal cloud moved toward it. Centered on the sun was another colored halo, circular or nearly so. And passing through the sun was a white circular halo.

The halo through the sun looks strange at first, but it is the easiest of the three halos to identify. It is the parhelic circle, whose radius must have been only 18°. What were the other halos?

Figure 6-10 is a simulation made with oriented plate crystals and randomly oriented crystals. Neither the parhelia nor the circumzenith arc are possible at this high sun elevation, but a new halo, the **circumhorizon arc**, appears in the lower part of the simulation. It is due to the oriented plates, which also made the parhelic circle, while the randomly oriented crystals made the 22° halo.

Figure 6-11, on the other hand, was made with only singly oriented columns. The halo around the sun is not the 22° halo but rather the circumscribed halo, which for high sun barely deviates from a circle. The weak arc near the bottom is the infralateral arc, unrecognizable from the low-sun infralateral arcs of Chapter 2. The faint halo intersecting the parhelic circle near the top is the Wegener arc.

Which of these two nearly identical simulations is correct? In the photograph the halo around the sun looks not quite circular, and its intensity falls off abruptly; it better matches the second simulation. However, the absence of the Wegener arc in the real display favors the first simulation, especially since tinkering with the crystal shapes and orientations in the second simulation cannot remove the Wegener arc and still leave a realistic parhelic circle and infralateral arc. Neither simulation is quite right.

Figure 6-12 is a compromise made with both oriented plates and singly oriented columns. The halos in it are the circumscribed halo, parhelic circle, and circumhorizon arc, the latter masking a faint infralateral arc. I like this simulation best.

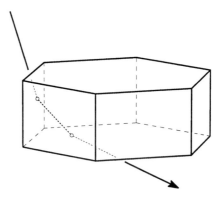

**Figure 6-13.** Ray path for the circumhorizon arc. The ray enters a prism face of an oriented plate and exits the bottom basal face.

### Display 6-7
### Barrow, Alaska, May 3, 1979

**Lower sunvex Parry arc.** Figure 6-15 shows a faint 22° halo, a lower tangent arc, and, just below it, the seldom photographed lower sunvex Parry arc. The halos were accompanied by an upper tangent arc and an upper Parry arc, out of the photograph. The display was seen from a small plane during a cloud seeding experiment.

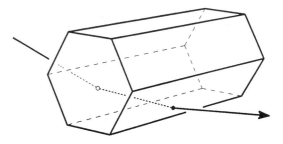

**Figure 6-14.** Ray path for the lower sunvex Parry arc. The ray enters one lower sloping prism face of a Parry oriented crystal and exits the other.

**Figure 6-15.** Lower tangent arc, lower sunvex Parry arc, and 22° halo. (Photograph by Takeshi Ohtake)

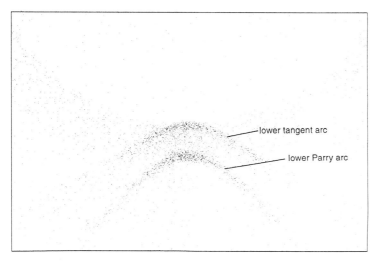

lower tangent arc

lower Parry arc

**Figure 6-16.** Simulation of the display. Singly oriented columns made the tangent arc, Parry oriented columns made the Parry arc, and randomly oriented crystals made the 22° halo. Sun elevation 32°.

**Display 6-8**

**Atacama Desert, Northern Chile, February 18, 1990**

**Lower suncave Parry arc.** Occurring only a day after Display 6-4, this is another fine high-sun display from the Atacama Desert. Unlike the earlier display, however, the brilliant circumscribed halo here is not alone. Note especially the Parry arcs — the upper and lower suncave Parry arcs. The lower one makes the circumscribed halo appear doubled at the bottom and then tails off faintly from the circumscribed halo toward the upper right. The upper one is positioned analogously, but nearly merges with the circumscribed halo on top.

The ray for the lower suncave Parry arc enters an upper sloping prism face of a Parry oriented crystal and exits the bottom prism face. As suggested by the ray path figure and confirmed by the simulations in Appendix C, the arc can only occur for fairly high sun. The sun elevation in this display was about 58°.

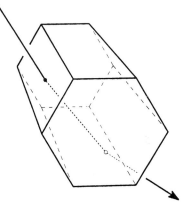

**Figure 6-17.** Ray path for the lower suncave Parry arc.

**Figure 6-18.** Bright circumscribed halo and parhelic circle, 22° halo, suncave Parry arcs, and faint Wegener arc. Display 6-8. (Copyright Gosewijn W. J. van Nieuwenhuizen)

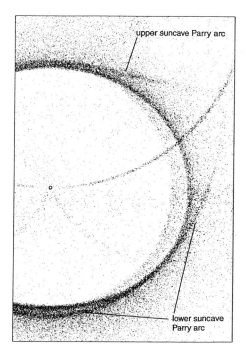

**Figure 6-19.** Simulation. Singly oriented columns, Parry oriented columns, and randomly oriented crystals made the halos.

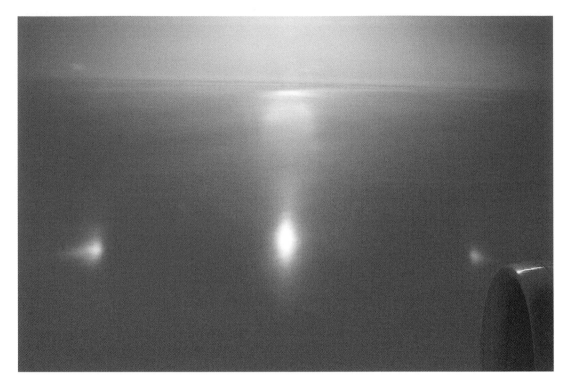

**Figure 7-1.** Lower tangent arc in upper center, subsun in lower center, and subparhelia. The sun is out of the photograph at the top. (Display 7-1)

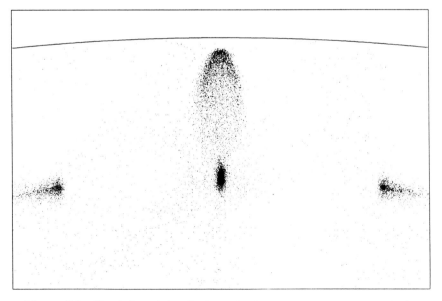

**Figure 7-2.** Simulation of the display. The lower tangent arc was made by singly oriented columns. The subsun and subparhelia were made by oriented plates, their tilts being only half a degree. Had there been crystals above the airplane, such small tilts would have made parhelia having almost no vertical extent, like the parhelion in Figure 1-19.

CHAPTER 7

SUBHORIZON HALOS

If you happen to be above ice crystals, you may see halos below the horizon.

## Display 7-1
## December 27, 1982

One way to have ice crystals below you is to fly. Display 7-1 was seen from a commercial jet flying between Seattle and Chicago. The photograph on the facing page, taken looking below the sun, shows the lower tangent arc and two new halos — the subsun and the subparhelia.

**Subsun.** The subsun is the bright white spot. It is directly below the sun and is as far below the horizon as the sun is above. Perhaps the simplest halo, the subsun arises in reflections from predominantly horizontal crystal faces. The faces act like a giant mirror, the subsun being the sun's reflected image. Because the faces need not be perfect to be effective, the subsun is one of the commonest halos. But of course you must be above the crystals to see it.

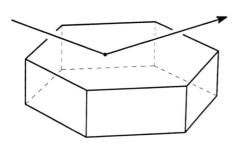

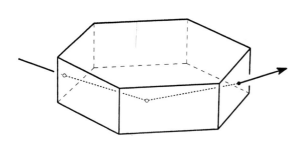

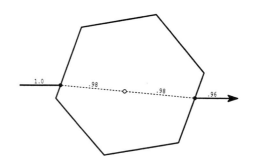

**Figure 7-3.** Two common ray paths for the subsun. The crystals here are oriented plates.
(Top) External reflection.
(Bottom) Internal reflection. The ray exits the prism face opposite the entry face.

**Subparhelia.** The two colored halos on either side of the subsun in Figure 7-1 are the subparhelia. They look like parhelia, but the parhelia would be above them and out of the photograph. The subparhelia are caused by oriented plates. Their ray paths are like those for the parhelia but with an extra internal reflection from the lower basal face.

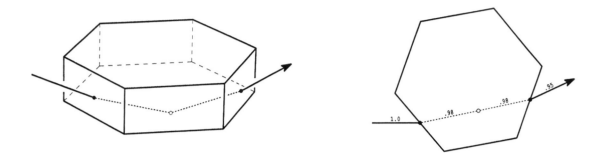

**Figure 7-4.** Common ray path for the left subparhelion.
Compare with the left parhelion ray path of Figure 1-7.

**Halos in the antisolar region.** The antisolar point on the celestial sphere is the antipodal point of the sun; you look toward the antisolar point to see the shadow of your head. The photograph in Figure 7-5, taken half a minute after Figure 7-1 and from the other side of the airplane, shows the region near the antisolar point. Two faint arcs pass diagonally through the antisolar point to form an X, and a brighter arc, the subparhelic circle, passes through the antisolar point horizontally.

A subparhelic circle might result from oriented plates (Figure 7-7 and Figure C-10) or from Parry oriented columns (Figure C-31). An X at the antisolar point might be either the antisolar arc, forming in Parry oriented columns (Figure C-31), or the diffuse B arcs, forming in singly oriented columns (Figure C-21). To sort out the possibilities, I ran several simulations of the display, the most successful being the one shown on the facing page. The crystals in the simulation were oriented plates and singly oriented columns, the latter having a slight tendency to maintain two prism faces horizontal. That is, the column crystals had poor Parry orientations, so the X in the simulation is a poorly defined antisolar arc or, equivalently, a cross between the antisolar arc and the diffuse arcs. The subparhelic circle was made by the oriented plates.

**Figure 7-5.** The display in the antisolar region. The white line at the top is the horizon. At the left is the airplane's engine.

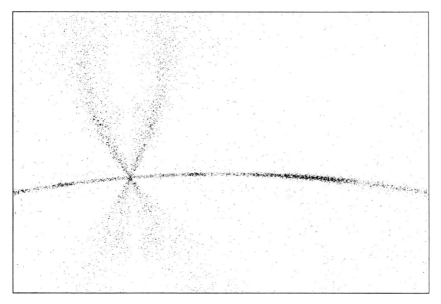

**Figure 7-6.** Simulation of Figure 7-5. Oriented plates made the subparhelic circle, and columns with poor Parry orientations made the X-shaped halo.

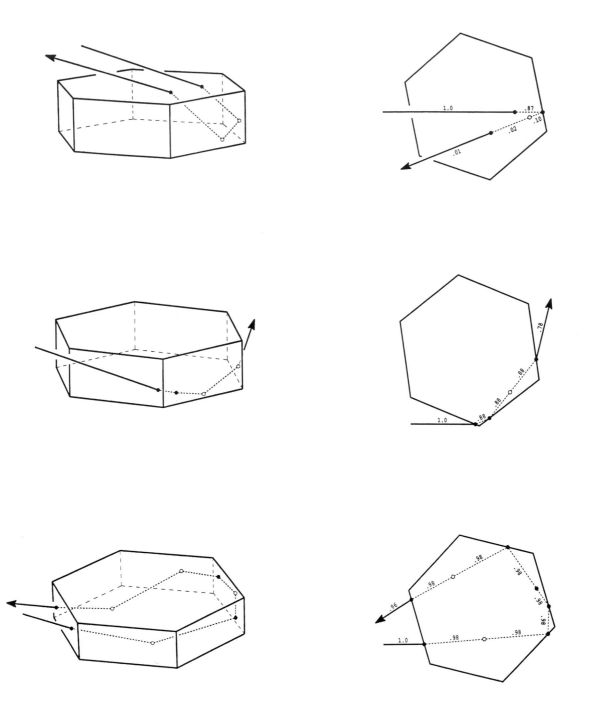

**Figure 7-7.** Three common subparhelic circle ray paths through oriented plates. The ray path in the bottom crystal makes a bright segment on the subparhelic circle, like the one in Figure 7-6 but to the left of the antisolar point; see Table 2 of Appendix E. Except for crystal shape and orientation, this ray path is essentially the diffuse A ray path of Figure 2-26.

## Display 7-2
## January 1, 1980

Here the sun elevation was 9.5° — compared with about 32° in Display 6-7 — and the lower (sunvex) Parry arc has assumed an inverted V shape. Also visible are a subsun and a faint subparhelion.

Simulations show that at this sun elevation the lower Parry arc and lower tangent arc look similar. Overexposing the photograph can make the halo look more like a lower tangent arc, so there is some question about its identification here.

**Figure 7-8.** Lower Parry arc seen from an airplane. (Display 7-2)

**Display 7-3**
**Eau Claire, Wisconsin, February 9, 1978**

In cold climates, you need not fly in order to see halos below the horizon. When tiny falling ice crystals are seen sparkling in the sunlight, a vantage point on a hill or building may reveal halos below. Figure 7-9 shows a subsun and subparhelion formed in crystals below the photographer.

When the subsun was introduced in Display 7-1 you may have noticed that it arises in the same way as the sun pillar (Display 6-1), namely, in reflections from more or less horizontal crystal faces. In fact, there is no sharp distinction between the two halos, but rather a continuous spectrum of halos. At one extreme is a truly circular image of the sun formed when the crystal faces are perfectly horizontal, and at the other, tall pillars formed when the faces are substantially tilted (assuming the sun is low). The elongated subsun shown below falls somewhere in between the two extremes.

**Figure 7-9.** Subsun and left subparhelion.

### Display 7-4
### Fairbanks, Alaska, January 30, 1991

The photograph below, taken from a rooftop, shows a nearly vertical shaft of light descending from the parhelion. What is it?

My first impression is that this halo is a secondary pillar. That is, the parhelion, instead of the sun, is acting as a light source and producing its own pillar. But then the parhelion should be much brighter than the pillar, just as the sun is brighter than an ordinary pillar.

The accompanying simulation, like all those in this book, involves only single scattering — no ray from the sun encounters more than one crystal. So the simulation cannot make secondary halos, yet it satisfactorily reproduces the display. Evidently the vertical halo need not be a secondary halo. In fact, the halo in the simulation turns out to be part parhelion and part subparhelion: The crystals in the simulation are oriented plates and randomly oriented plates, and computer examination of the ray paths responsible for the vertical halo reveals only parhelion and subparhelion ray paths.

This vertical halo has been photographed from an airplane [Können, 1985, p. 59], and there are suggestions of it in some Antarctic displays, yet it is hardly mentioned in the literature. Why is it rare? To produce it, the sun must be low and the crystals not too thick, but I suspect the main reason for its scarcity is the need for crystals below or level with the observer. In the present display the halo is indeed largely below the horizon.

Surely displays like this one have prompted reports of "Lowitz arcs" (Chapter 11).

**Figure 7-11.** Simulation. Oriented plates and randomly oriented plates made the halos.

**Figure 7-10.** Unusual vertical halo below the parhelion. A fragment of the 22° halo is also visible. (Photograph by James F. Conner)

**Figure 8-1.**  Light pillars. Fairbanks, Alaska, March 1, 1983.

**Figure 8-2.**  A 22° halo formed in frost crystals on the ground. The finger is just to block the sun. Eau Claire, Wisconsin, 1978.

CHAPTER 8

COLD WEATHER HALOS

Halo formation requires ice crystals and a source of light. But the light source need not be the sun or moon, and the crystals need not be at normal cloud levels. Halos can therefore take strange forms during cold weather.

At subzero (Fahrenheit) temperatures, ice crystals may form at low levels in the atmosphere. Halos may then appear in front of nearby buildings or even below the horizon, as we have seen. At night, streetlights can produce halos. Streetlight pillars are especially conspicuous, but all of the common halos can be seen in streetlights if an observer is persistent. The search for streetlight halos is well worth making, because most of them are strikingly three-dimensional, due to the proximity of the light source. Unfortunately, a single camera by itself cannot capture their spatial character.

The ice crystals that cause halos need not even be falling through the atmosphere. Low sun on crisp autumn mornings can make striking 22° and 46° halos in frost on the ground; look for them on well-kept athletic fields, for example. In winter the 22° and 46° halos can be seen occasionally in ice crystals on snow surfaces, especially at sub-zero temperatures. These surface halos look hyperbolic if you fix attention on the ground, but circular if you unfocus your eyes slightly.

I once saw parhelia (subparhelia?) on a snow surface. More common are subparhelia formed in ice crystals on car roofs.

Because we tend to see only what we look for, many common phenomena go unnoticed. The halos described above are no exception. In cold weather you will see them often, once you become aware of them, and you will wonder how you ever missed them.

**Figure 8-3.** Right parhelion formed in nearby crystals. Fairbanks, Alaska, March 10, 1990.

CHAPTER 9

ORGANIZING THE HALOS

The number of halos is intimidating. How can the halo menagerie be tamed?

Different halos result from different crystal orientations or from different ray paths through the crystals. Either theme — crystal orientations or ray paths — can be used to group halos and thus to make the subject manageable.

In the first four chapters, halos were grouped by means of the crystal orientations that caused them. Oriented plates caused the halos of Chapter 1, singly oriented columns caused the halos of Chapter 2, Parry oriented columns caused the halos of Chapter 3, and randomly oriented crystals caused the circular halos of Chapter 4, or so I will assume now.

In the present chapter, halos are grouped by means of their ray paths. Consider, for example, all halos whose ray paths, regardless of crystal orientation, are like parhelion ray paths, with the ray entering a prism face and exiting an alternate prism face.[1] These halos would be the 22° halo, tangent arcs, Parry arcs, and parhelia, all shown in the simulations in Figure 9-1.[2] To emphasize the new perspective, compare these simulations with, say, Figure 3-1. There the crystal orientations were all of one kind, and different halos arose from different ray paths. In Figure 9-1 the ray paths are all of one kind, and different halos arise from different crystal orientations.

Thus the 22° halo, tangent arcs, Parry arcs, and parhelia are closely related. Of these, the 22° halo is the most general. It arises in random — that is, all — crystal orientations, whereas the other halos — tangent arcs, Parry arcs, and parhelia — arise in restricted subsets of orientations. These other halos therefore lie within the 22° halo. Similarly, the Parry arcs lie within the tangent arcs, because Parry orientations are a subset of singly oriented column orientations.

Specifying a ray path — the parhelion ray path — has determined a family of related halos. Other ray paths determine other families of halos. Figure 9-2 shows the halos whose ray paths are like 46° halo ray paths, with the ray entering a prism face and directly exiting a basal face, or vice versa. Figure 9-3 shows the halos whose ray paths are like subparhelion ray paths.

You may enjoy thinking about the halo families determined by other ray paths. What halos would you expect from the subsun ray paths of Figure 7-3? What halos would you expect from 120° parhelion ray paths?

Grouping by ray path leads to complications if a ray path lights only part of a halo. Figure 9-4 shows some halos whose ray paths are like the diffuse arc ray path at the top in Figure 2-26. The "halos" are the diffuse A arcs, together with fragments of the antisolar arc and subparhelic circle. Other ray paths would be needed to flesh out the antisolar arc and subparhelic circle.

Of course, I never expect to see a display like Figure 9-4, since I cannot imagine why Nature would only allow ray paths like those for the diffuse A arcs. The simulations in this chapter were obviously not created to reproduce real displays.

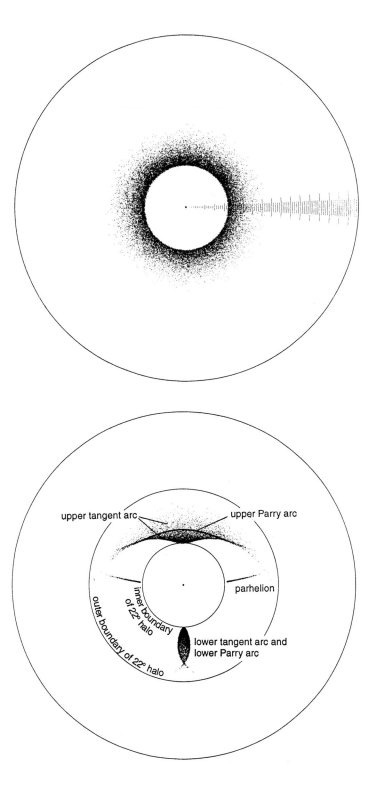

**Figure 9-1.** Family of related halos whose ray paths are like parhelion ray paths.

(Top) The 22° halo, made by randomly oriented crystals. The halo is an annular region with inner boundary 22° from the sun. In principle the halo has an outer boundary as well, 50° from the sun, but it is exceedingly weak there. The larger tick marks occur every five degrees.

(Bottom) Tangent arcs, made by singly oriented columns; Parry arcs, made by Parry oriented columns; and parhelia, made by oriented plates. The halos lie within the annular region occupied by the 22° halo. The upper tangent arc is a two-dimensional region that extends upward from its bright lower boundary to include the dots above. The upper Parry arc is the bright curve within the upper tangent arc. Similar remarks apply to the lower tangent arc and lower Parry arc, but the lower Parry arc here nearly coincides with the bright boundary of the lower tangent arc and is difficult to distinguish. The simulations are centered on the sun, and the lower tangent arc would be just below the horizon.

All simulations in this chapter are fisheye views. The sun elevation is 20° for each.

---

1 Any even number of internal basal face reflections is also allowed. In Appendix E terminology, the ray paths are subtype *35* or *37*.

2 Lowitz arcs (Chapter 11), if due to spinning plate crystals, should also be included here. See Tape [1980].

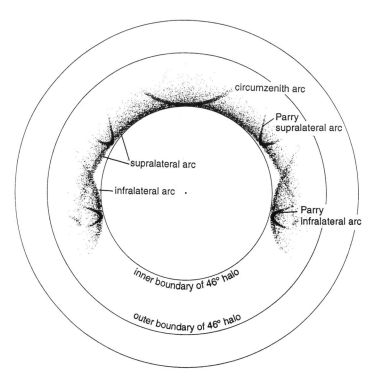

**Figure 9-2.** Halos whose ray paths are like 46° halo ray paths: the 46° halo, made by randomly oriented crystals; supralateral and infralateral arcs, made by singly oriented columns; Parry supralateral and Parry infralateral arcs, made by Parry oriented columns; and circumzenith arc, made by Parry oriented columns or oriented plates. The 46° halo is indicated only by its inner and outer boundaries. The supralateral and infralateral arcs are two-dimensional regions that lie within the 46° halo. The Parry supralateral arcs and the circumzenith arc in turn lie within the supralateral arc, and the Parry infralateral arcs lie within the infralateral arcs.

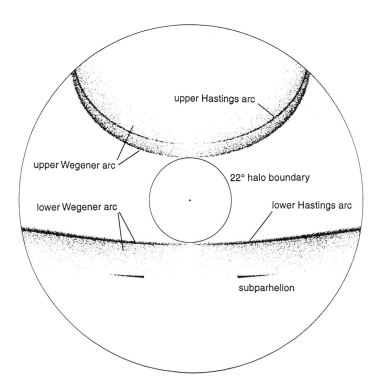

**Figure 9-3.** Halos whose ray paths are like subparhelion ray paths: Wegener arcs, made by singly oriented columns; Hastings arcs, made by Parry oriented columns; and subparhelia, made by oriented plates. The upper Wegener arc is a two-dimensional region, and the upper Hastings arc is a curve within it. Similar remarks apply to the lower Wegener arc and lower Hastings arc, but the lower Hastings arc here nearly coincides with the bright boundary of the lower Wegener arc and is difficult to distinguish. The "halo" caused by randomly oriented crystals is an enormous and diffuse affair, indicated here only by its inner boundary, which coincides with that of the 22° halo. Figure 9-3 and Figure 9-1 correspond closely, since only a basal face reflection separates their halos.

**Figure 9-4.** Some "halos" whose ray paths are like ray paths for the diffuse A arcs. The view is looking away from the sun, the horizontal line being the horizon.

(Right) Diffuse A arcs, made by singly oriented columns.

(Bottom left) Segments of the antisolar arc and subparhelic circle, made by Parry oriented columns. These segments must lie within the diffuse A arcs, though certainly not within the bright part.

(Bottom right) Segments of the subparhelic circle, made by oriented plates.

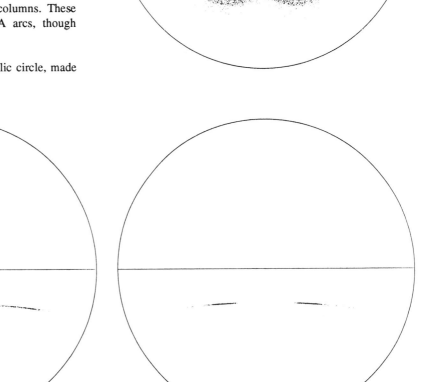

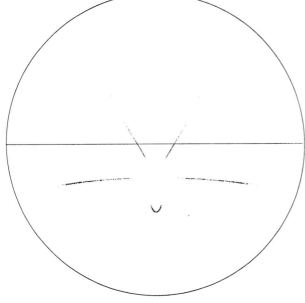

CHAPTER 10

## PYRAMIDAL CRYSTALS AND ODD RADIUS CIRCULAR HALOS

Some ice crystals have pyramidal faces, that is, faces meeting the crystal axis obliquely. Such crystals I will call pyramidal, whether or not they also have prism and basal faces. Pyramidal crystals are not rare — a crystal sample will often contain a few — but they seldom occur in quantity. When they do, they can produce rare circular halos having unusual radii — neither 22° nor 46°. These circular halos, of course, result when the crystals orient more or less randomly as they fall. If the crystals orient preferentially, they cause intensity variations along the halos, or even additional exotic arcs.

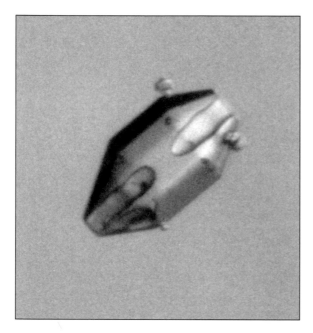

**Figure 10-1.** Two large pyramidal ice crystals collected during a light snow-fall. The upper crystal, about 0.18 mm in length, appears to have six prism faces, two severely indented basal faces, and twelve pyramidal faces. The lower crystal has six prism faces, two basal faces, and six pyramidal faces. The tiny objects adhering to the crystals are probably ice fog crystals. Fairbanks, Alaska, January 27, 1989. Temperature -39° F.

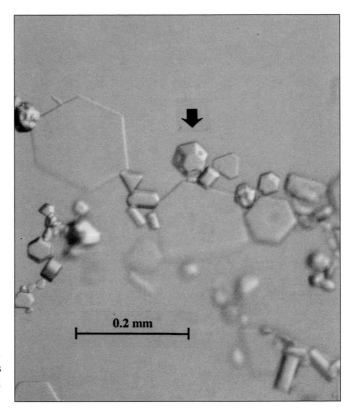

**Figure 10-2.** Pyramidal crystal (arrow). Its hexagonal face is a basal face, and its adjacent trapezoidal faces are pyramidal faces. South Pole, January 31, 1990.

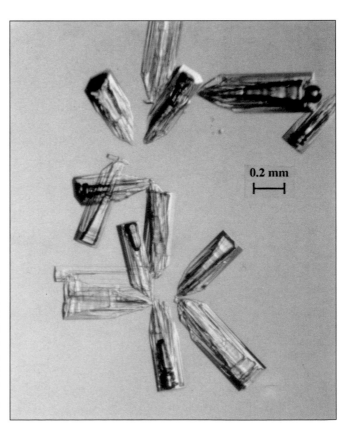

**Figure 10-3.** Bullet crystals — poorly formed columns with one end tapered. Bullet crystals have sometimes been confused with pyramidal crystals, but their tapered ends do not have pyramidal faces and do not cause odd radius halos. Bullet crystals are common and are responsible for some mediocre 22° halos and tangent arcs. Fairbanks, Alaska, January 17, 1989.

**Figure 10-4.** Odd radius halo display. (Display 10-1)

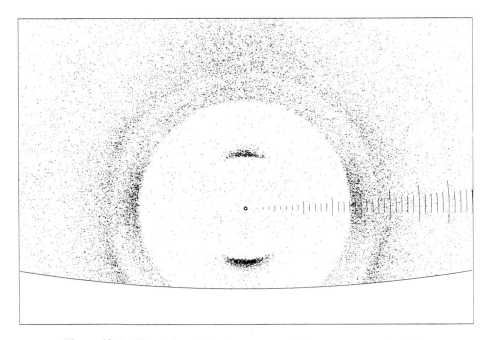

**Figure 10-5.** Simulation of the display. Pyramidal crystals made the halos. The tick marks, at 1° intervals, show that none of these halos is the ordinary 22° halo.

### Display 10-1
### Fairbanks, Alaska, March 7, 1989

The photograph on the facing page shows two fairly bright circular halos having measured radii of about 18.1° and 22.4°.[1] Much closer to the sun is a less conspicuous halo, stronger directly above the sun but largely lost in the glare. Intensity variations along the halos, perhaps indicative of additional halos, complicate the picture.

This seems to be the first good display of odd radius halos during which crystals were sampled.[2] More than half the sample turned out to consist of pyramidal crystals. Their predominance here, together with their scarcity during ordinary halo displays, suggests that pyramidal crystals are indeed the cause of odd radius halos.

In size and shape, the few crystals shown below are typical of those collected, the prism faces often being small or absent. Similar crystal shapes, shown in Figure 10-7, were used to simulate the display. The resulting simulation, Figure 10-5, seems adequate: The halo radii are consistent with the measured values, and the intensity variations match those in the photograph. Note especially the brighter regions at the 10:00 and 2:00 positions about 24° from the sun, and at 9:00 and 3:00 about 18° from the sun.

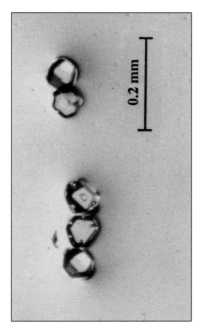
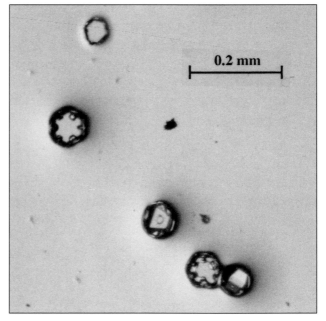

**Figure 10-6.** Plastic replicas of ice crystals from the display. Nearly all those shown are pyramidal. To make the replicas, a glass slide was coated with acrylic spray and then waved through the air to intercept the crystals. The slide thus contains thousands of crystal replicas, only a few of which are shown. The special atmospheric conditions that produced these pyramidal crystals are unknown. The crystals fell from a power plant plume, the air temperature at the ground being about -10° F.

---

1 These measured radii are uncertain by about 0.75°. They are obtained from the photograph by careful comparisons with a photograph of a star field taken with the same lens. The main source of error is the fuzziness of the halo edge.

2 Earlier, Steinmetz and Weickmann [1947] had seen a fine display of odd radius halos from an airplane. They sampled crystals during the flight but apparently lacked a microscope for systematic examination.

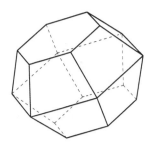

**Figure 10-7.** Crystal shapes used in making the simulation in Figure 10-5. Crystals like that at the left were randomly oriented. Crystals like that at the right were given some tendency to fall with their small basal faces downward, as shown, and they are responsible for intensity variations along the halos. The ray path shown is for a circular halo of radius 18.3°.

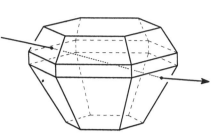

**Crystal orientations and ray paths for the circular halos.** This and similar simulations suggest that the crystals that cause the odd radius circular halos, like those that cause the ordinary 22° and 46° circular halos, are more or less randomly oriented. The ray paths, again like those for the ordinary circular halos, proceed directly from an entry face to an exit face, with internal reflections playing no essential role. What distinguishes among all these circular halos is the angle between the entry face and exit face. This angle, in fact, can be used to calculate the halo radius.[3] An interfacial angle of 60°, for example, gives the ordinary 22° halo, while an angle of 90° gives the 46° halo, and an angle of 52.4°, which occurs in the figure above, gives a halo of radius 18.3°.

**The pyramidal faces are {1 0 -1 1} (simple).** The relation between interfacial angle and halo radius was well known in nineteenth century halo theory; change the angle, change the halo radius. Almost any circular halo could therefore be "explained" by postulating crystals with pyramidal faces suitably inclined to one another. The explanation was deficient for several reasons, but especially because it was offered with little assurance that the postulated shapes actually existed. Indeed, hardly anyone had ever seen a pyramidal ice crystal.

But by the early twentieth century, X-ray diffraction had revealed the crystallographic structure of ice. This structure limits the theoretically possible crystal faces, in the sense that the angles between faces can assume only certain discrete values. The simplest theoretically possible pyramidal faces are the so-called {1 0 -1 1} faces. They are supposed to make an angle of 28° with the crystal axis.[4]

---

3 Assuming that the ray within the crystal proceeds directly from entry to exit face (or that internal reflections cause no net change in the ray's direction), one finds from Snell's law that $\sin I = n \sin(A/2)$ and $D = -A + 2 I$, where $n = 1.31$ is the refractive index of ice (yellow light), A is the angle between entry and exit faces, I is the angle of incidence giving minimum deviation, and D is the minimum deviation, equal to the halo radius.

In calculating halo radius, one might think that some subtraction should be made for the angular size of the sun and for the variation of refractive index with wavelength. However, simulations show that while these factors blur the halo's inner edge, the subjective impression is not one of a decrease in radius.

For values of A larger than 99.5°, which corresponds to a halo radius of 80.5°, the first equation above cannot be solved — an indication of total internal reflection at the proposed exit face. Larger circular halos are therefore impossible. Simulations show that circular halos having radii larger than about 50° are probably too weak to be seen.

4 The symbols {1 0 -1 1} are the Miller indices of the faces. For a definition of Miller indices and an indication of why the {1 0 -1 1} faces should be considered simple, see Mason and Berry [1968]. The value 28° is calculated using a crystallographic axial ratio c/a = 1.63, which is the ratio determined by X-ray diffraction [Hobbs, 1974].

Most pyramidal faces look like they could be {1 0 -1 1} faces, but verifying that they are inclined 28° to the axis is usually difficult. The photographs in Figure 10-1, however, are good enough that the relevant angles can be measured, and the results turn out to be within a degree of the expected values.[5] Thus the {1 0 -1 1} faces apparently do occur in real ice crystals. Crystal observations made by Kobayashi and Higuchi [1957] and by Kobayashi [1965] point to the same conclusion.

All pyramidal faces used in simulations in this chapter are {1 0 -1 1} faces. Their simplicity and their known occurrence help to justify their use. And avoiding the use of other pyramidal faces makes the simulations more convincing, if, of course, the {1 0 -1 1} faces can reproduce the observed halos.

**The circular halos arising from the {1 0 -1 1} faces.** Since the radius of a circular halo depends on the angle between the entry and exit faces of its ray path, the number of circular halos is limited by the number of values for the interfacial angles on the crystals. If the only pyramidal faces are {1 0 -1 1} faces, then the circular halos are limited to those listed in the table below.

**The halos in the display.** Not all the halos in the table were present in Display 10-1. Computer examination of the ray paths involved in the simulation reveals ray paths mainly for the 9°, 18.3°, 19.9°, 22.9°, and 23.8° halos. In the simulation and presumably in the display itself, the halo at about 18° is a combination of the 18.3° and 19.9° halos, which overlap and are difficult to distinguish. Similarly, the halo at about 23° is a combination of the 22.9° and 23.8° halos. The bright regions at 10:00 and 2:00 are due to 23.8° halo ray paths in preferentially oriented crystals, while those at 9:00 and 3:00 are due to 18.3° ray paths. The diffuse illumination near the top is due to 22.9° ray paths.

If this analysis is correct, then neither of the bright halos in the photograph is the ordinary 22° halo.

| Entry and exit faces | | Interfacial angle | Theoretical halo radius | Observed halo radius |
|---|---|---|---|---|
| 3-26 | Figure 10-14 | 28.0° | 9.0° | 9.1° |
| 13-25 | Figure 10-7 | 52.4 | 18.3 | 18.0 |
| 23-26 | | 56.0 | 19.9 | 20.1 |
| 3-5 | Chapter 4 | 60.0 | 21.8 | 22.0 |
| 1-25 | | 62.0 | 22.9 | ... |
| 3-25 | Figure 10-20 | 63.8 | 23.8 | 23.6 |
| 23-25 | | 80.2 | 34.9 | 35.5 |
| 1-5 | Chapter 4 | 90.0 | 45.7 | ... |

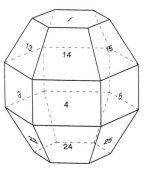

**Figure 10-8.** Table showing the theoretically possible circular halos arising in crystals having prism faces, basal faces, and {1 0 -1 1} pyramidal faces. The last column gives the halo radii measured by Neiman in Display 10-2 (next section). The crystal diagram shows the numbering of faces used in the table. Basal faces are 1 and 2, prism faces are 3 through 8, top pyramidal faces are 13 through 18, and bottom pyramidal faces are 23 through 28.

---

5 If you measure the angles using Figure 10-1, note that each crystal is probably lying on a prism face and that you are measuring the angle between a crystal edge — not a face — and the crystal axis. The theoretically expected value is 31.5°.

### Display 10-2
### Boulder, Colorado, July 21, 1986

This elaborate display had at least six circular halos. The observer, Paul J. Neiman, measured the halo radii from his photographs and found values close to those expected from crystals with {1 0 -1 1} faces, as shown in the table in Figure 10-8 [Neiman, 1989]. In fact, all of the circular halos whose ray paths involve {1 0 -1 1} faces and prism faces were apparently present in the display.

To simulate the display, I started with pyramidal crystals shaped like the one below. In Figure 10-12 the crystals were randomly oriented and produced five distinguishable circular halos: the 9°, 18.3°, 19.9°, 23.8°, and 34.9° halos in the table. Then in Figure 10-13 the same pyramidal crystals were given some tendency to maintain their axes horizontal. The resulting intensity variations along the halos match the photographs better than the uniform halos in the first simulation. Also, some prismatic columnar crystals were added in the second simulation to make the ordinary 22° halo (really a poor circumscribed halo), which is absent in the first simulation. Although no crystal sample is available for this display, the striking agreement between the simulation and the photographs makes the simulation crystals seem plausible candidates for the real crystals.

**Figure 10-9.** Odd radius halo display; Display 10-2. Notice the faint outer halo. (Copyright Paul J. Neiman)

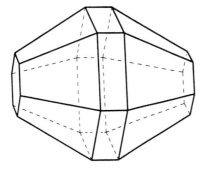

**Figure 10-10.** Shape of pyramidal crystals used in Figure 10-12 and Figure 10-13. This crystal is identical to the crystals in Figure 10-7 (right), Figure 10-14, and Figure 10-20, in the sense that the faces are prism faces, basal faces, and {1 0 -1 1} pyramidal faces.

**Figure 10-11.** Display 10-2 photographed with a longer lens. Notice the two faint halos just inside the bright outer halo. (Copyright Paul J. Neiman)

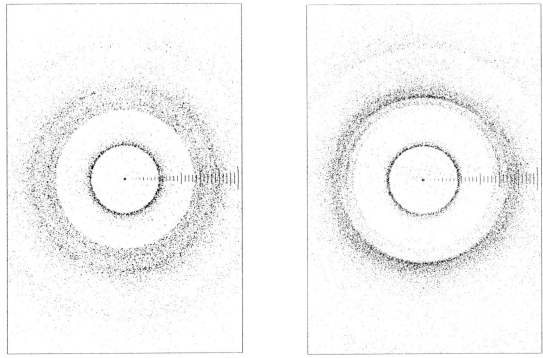

**Figure 10-12.** (Left) Simulation of the display using randomly oriented pyramidal crystals. The 9, 18.3, 19.9, 23.8, and 34.9° halos of Figure 10-8 are discernible. **Figure 10-13.** (Right) Same as Figure 10-12 except that here the pyramidal crystals were given some tendency to maintain their axes horizontal, and some ordinary singly oriented columns were added. Notice the weakness of the 18.3° halo at the top and the absence of the 19.9° halo at the bottom, in agreement with the photographs.

## Display 10-3
### Seattle, Washington, May 1, 1973

The two halos in the photograph at the right are probably the 9° halo and the usual 22° halo. To simulate them, I used randomly oriented pyramidal crystals like the one below, together with singly oriented ordinary prismatic columns. In the resulting simulation, the pyramidal crystals made the 9° and 22° halos, and the ordinary columns made the enhancements of the 22° halo at top and bottom.

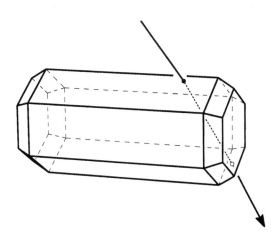

**Figure 10-14.** Shape of pyramidal crystals used in making Figure 10-16. Some ordinary prismatic columns were used as well. The ray path is for the 9° halo.

**Relative intensities of odd radius halos.** As already explained, the crystals used in making the simulations in this chapter have only prism faces, basal faces, and {1 0 -1 1} pyramidal faces — no other pyramidal faces. A surprising variety of simulations can nevertheless be produced from such crystals, even when the crystals are randomly oriented. The resulting halos are limited to those in the table of Figure 10-8, but their intensities can vary enormously, to the point where some halos may even disappear. The 18.3° halo, for example, is strong in Figure 10-5 but absent in Figure 10-16. (Probably its ray path is incompatible with the long prism faces in Figure 10-14; compare Figure 10-7, which shows the ray path.) Changing the crystal shapes — without, of course, changing angles between faces — can drastically change the halo intensities.

By choosing suitable crystal shapes and perhaps by giving the crystals preferred orientations, but without resorting to pyramidal faces other than {1 0 -1 1}, one can successfully simulate most existing photographs of odd radius halos.[6] The next display is another impressive example.

---

6 Exceptions may be "elliptical halos" — small oblong features that have been seen around the sun or moon. The origin of these rare and transient phenomena is unknown. See Hakumäki and Pekkola [1989].

**Figure 10-15.** Odd radius halo display; Display 10-3. The display is described by Turner and Radke [1975]. (Photograph by Francis M. Turner)

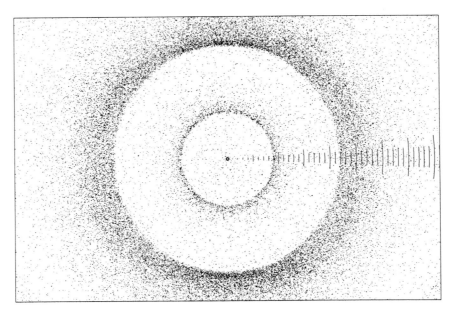

**Figure 10-16.** Simulation of the display. Randomly oriented pyramidal crystals and singly oriented columns made the halos.

### Display 10-4
### Georg von Neumayer Station, Antarctica, July 11, 1987

Dominating this complex display are familiar halos from ordinary prismatic crystals, especially the 22° halo and the upper and lower tangent arcs (Figure 10-17). But it is the other halos that should by now catch your eye: the 9° halo and, especially, several more or less concentrated patches of light in unexpected places. These patches of light are "parhelia" of odd radius halos. Suggestions of such parhelia appeared in Display 10-1, but here they are much stronger.

**Parhelia of the odd radius halos.** By a "parhelion" of a circular halo, I mean any halo arising in crystals and ray paths like those for the circular halo, but with the crystals having preferred orientations. In this sense of "parhelion," the parhelia of the ordinary 22° halo would be the upper and lower tangent arcs and the Parry arcs, as well as the ordinary parhelia. Similarly, each odd radius halo in principle has a variety of parhelia.

The simulation in Figure 10-18 was made using pyramidal crystals shaped like the one in Figure 10-20. The crystals had their basal faces almost perfectly horizontal, with the smaller basal face downward as shown, and produced sharply defined odd radius parhelia. Then in Figure 10-19 these same crystals were allowed to deviate more from the horizontal. These less constrained crystals produced parhelia that are less well defined, more like those in the photograph. Randomly oriented pyramidal crystals shaped like the

**Figure 10-17.** Halo display with odd radius parhelia. This is a lunar display, and the short white streaks are made by stars during the time exposure. The faint thin circle passing nearly through the moon is not real but rather an artifact of the camera. (Photograph by Klaus Sturm)

one in Figure 10-14 were added to make the 9° circular halo, and ordinary prismatic crystals were added to make the remaining halos. If this interpretation is correct, then in this display we are seeing parhelia of the 9°, 18.3°, 22.9°, 23.8°, and 34.9° odd radius halos: The 9° and 23.8° parhelia are clear. The 22.9° parhelion is also clear but apt to be mistaken for a Parry arc. The 18.3° parhelia are faint, and the 34.9° parhelia are diffuse. Presumably there is also a 19.9° parhelion, but masked by the lower tangent arc.

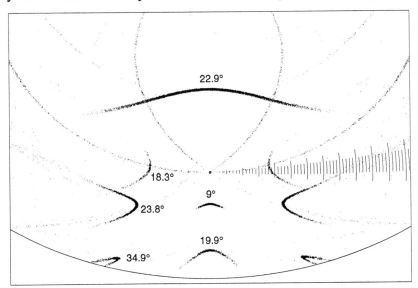

**Figure 10-18.** Simulation made with pyramidal crystals shaped and oriented as in Figure 10-20. Most of the halos are parhelia of odd radius circular halos. The number next to each parhelion gives the radius of the corresponding circular halo (Figure 10-8). Since the parhelion need not be tangent to its circular halo, the number need not be the exact angular distance from the parhelion to the sun.

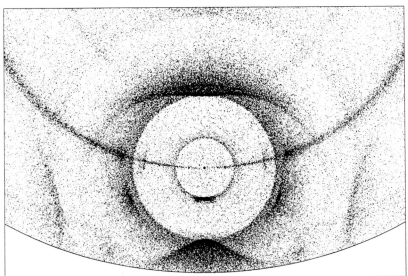

**Figure 10-19.** Simulation of the display. Pyramidal crystals like the one in Figure 10-20 made the odd radius parhelia, which are poorly defined versions of those in Figure 10-18. Randomly oriented pyramidal crystals like the one in Figure 10-14 made the 9° circular halo and part of the 22° halo intensity. Ordinary prismatic columns and plates made the remaining, familiar halos.

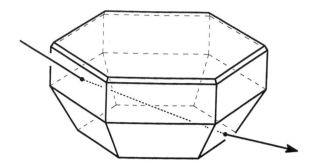

**Figure 10-20.** Shape of crystals that made the odd radius parhelia in Figure 10-18 and Figure 10-19. The ray path is for the 23.8° halo.

**Frequency of odd radius halos.** In Antarctica I often saw odd radius parhelia, usually unaccompanied by the corresponding circular halos. The parhelia were similar to those in Display 10-4 but weaker. Strongest and most frequent was the 9° parhelion, but always close to the sun, of course, and therefore easily missed. Considerably weaker were the 22.9° and 23.8° parhelia, in fact usually so weak and poorly defined that they too could easily be missed.

In Fairbanks each winter I normally see a handful of odd radius halo displays in low-level crystals. The best was Display 10-1, with most displays being much weaker. Odd radius displays at normal cloud levels, like Display 10-2 and Display 10-3, are seen far less often, at least from the ground. Limited evidence suggests that observing from aircraft is more productive.

**Other odd radius halos.** Again I stress that in the simulations in this chapter the pyramidal faces are only the {1 0 -1 1} faces. As already mentioned, these faces, although they cannot produce circular halos different from those in the table in Figure 10-8, are enough to simulate most existing photographs of odd radius halo displays.

There do exist reports, largely unsubstantiated by photographs, of circular halos different from those in the table [Visser, 1942-1961]. Such halos would require crystals having pyramidal faces other than {1 0 -1 1} faces. I have seen a few such crystals; perhaps on rare occasions they occur in sufficient quantity to produce halos.

More photographs of odd radius halos are obviously needed. The halos are rare, but not so rare as has often been supposed, and the patient skywatcher will see them eventually.

# CHAPTER 11

## HEVEL'S HALO AND OTHER MYSTERIES

Scattered through the scientific literature are reports of unexplained halos. Most of the accounts are obviously unreliable and justifiably forgotten, but a few persist, some over centuries, resisting attempts to explain them away.

### Display 11-1 — The Danzig Display
### Gdansk (Danzig), February 20, 1661

This display, known as the Danzig Display or Seven Suns Display, was reported by the famous seventeenth century astronomer Johannes Hevelius in his beautiful but rare *Mercurius in Sole visus Gedani*,... The book consists mostly of astronomical observations, including the transit of Mercury referred to in the title, but in four pages at the end, Hevelius carefully describes the halo display. Figure 11-2 is a diagram of the display, taken from the book.

**Hevel's halo.** Most of the halos in Hevelius's diagram look familiar: the 22° halo, parhelia, circumzenith arc, parhelic circle, upper and lower tangent arcs, anthelion, and probably infralateral and supralateral arcs. But what about the arc *NEKDP*? Now known as Hevel's halo, it is supposed to be a 90° circular halo centered on the sun. In the centuries since Hevelius's observation, no one has found a mechanism to produce such a halo.[1] If Hevelius's description is correct, then halo theory is deficient. Could he have made a mistake?

**Figure 11-1.** Hevelius's large azimuthal quadrant. (From Hevelius's *Machinae Coelestis Pars Prior*)

---

1 One mechanism often mentioned involves, in Appendix E terminology, *3573* ray paths [e.g., Visser, 1942-1961]. Another involves pyramidal crystals with {2 2 -4 3} faces [Humphreys, 1940]. I have used simulations to test these and other mechanisms. None produced a perceptible 90° halo.

**The case against Hevel's halo.** Halo reporting without a camera is not easy, especially when the display is complex or changing. Time for notetaking is short, and recollections without notes are unreliable. Estimates of large angular distances in the sky are especially prone to error, and, indeed, Hevelius's account contains inconsistencies that weaken his angular measurements.[2] So perhaps the halo seen by Hevelius was not really a 90° circular halo at all.

Perhaps in fact the halo was the subhelic arc [Greenler, 1980]. Both *NEKDP* and the subhelic arc are large halos in the same general region of sky, and both consist of two separate white arcs. Hevelius incorrectly reported the upper tangent arc *QGR* as a 45° circular arc centered on the zenith; this predilection for simple, circular geometry might have skewed his description of *NEKDP* as well. Perhaps Hevelius simply did not get the subhelic arc quite right.

I know of no photographs of a 90° circular halo. A few reports of it do exist,[3] but most are less convincing than Hevelius's own. One is tempted to conclude, along with Hastings [1902], that "an inexplicable phenomenon recorded only once in a quarter of a millennium does not really exist."

**The case for Hevel's halo.** On the other hand, Hevelius was apparently an experienced halo observer. Except for the halo in question, his drawing looks about right, and so do his drawings of other halo displays. And he certainly knew his way around the celestial sphere. His observatory, the best in Europe at the time, spanned three rooftops and housed huge quadrants, sextants, and octants that he had used for years in precise determinations of stellar and planetary positions. When Hevelius said the halo radius was about 90°, he could not have been far wrong.

What about the conjecture that *NEKDP* was the subhelic arc? I once thought so, but doubts arose as I learned more about Hevelius and as I studied his (difficult Latin) description of the display. The simulation at the right shows the disparity between the subhelic arc and a 90° halo. Hevelius reported that the halo in question reached the horizon, but a subhelic arc reaching the horizon is far from a 90° halo, as the simulation shows. He also stated that the halo crossed the ecliptic at right angles, as a 90° halo must; the subhelic arc does not. Finally, a subhelic arc should have been accompanied by a Wegener arc, as in the simulation, but no Wegener arc was mentioned.

Several strange old halo drawings have been subsequently substantiated by photographs of similar halos. Perhaps Hevelius's diagram will be another. In general, we perhaps too easily dismiss the work of previous generations.

I am torn between the opposing arguments and can only offer the obvious: A suitable photograph or theoretical explanation would vindicate Hevelius. In the meantime, the case for Hevel's halo continues to weaken.

---

2 A 90° halo centered on the sun would meet the horizon at points 90° in azimuth from the sun. It would cross the parhelic circle at points greater than 90° in azimuth from the sun. Arc *NEKDP* fails on both counts. However, it does pass through the pole of the ecliptic, as a 90° halo must.

3 See, for example, Besson [1914] and Pernter and Exner [1922]. Also see Display 11-4.

Only recently has the subhelic arc been widely recognized as theoretically possible. Ignorance of it has surely helped to legitimize Hevel's halo. Whatever the truth about the halo seen by Hevelius himself, subsequent observers must have mistaken some subhelic arcs for Hevel's halo, simply because there was no other candidate available.

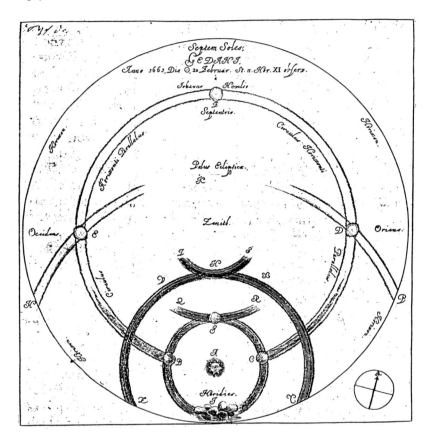

**Figure 11-2.** Hevelius's diagram of the Danzig Display. The arc *NEKDP* is Hevel's halo. This copper plate engraving was probably made by Hevelius himself.

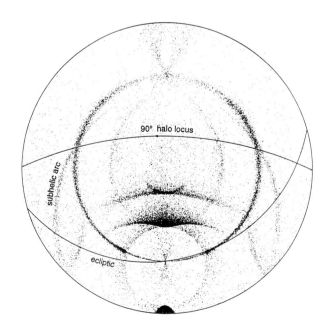

**Figure 11-3.** An attempt to simulate Hevelius's diagram. Singly oriented columns, oriented plates, and randomly oriented crystals made the halos. This simulation uses stereographic projection to depict the halos. Circles on the celestial sphere are therefore represented as true circles in the simulation, and angles on the celestial sphere are represented correctly.

## Display 11-2 — The St. Petersburg Display
### St. Petersburg, Russia, June 18, 1790

Often cited as a complicated and rare display, this is perhaps the most famous of all halo displays. Its chronicler was one Tobias Lowitz, who made the drawing reproduced at the right. Actually, the drawing shows the display as a whole to be no more elaborate than, say, Display 2-4, and today most of its halos are well understood.

**Lowitz arcs.** What remains special about the display are two short colored arcs extending from the parhelia inward and downward to the 22° halo. Now known as Lowitz arcs, they remain controversial. What exactly did Lowitz see, and what was the cause?

The conventional answer is that Lowitz arcs arise in spinning plate crystals. In each crystal a long axis of symmetry is supposed to remain horizontal while the crystal spins about it. The simulation in Figure 11-6, which was made with spinning plates as well as more familiar crystal orientations, nicely reproduces the Lowitz arcs in the drawing.

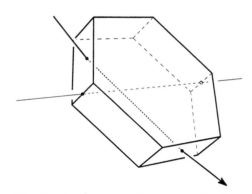

**Figure 11-4.** Spinning plate crystal and ray path for the Lowitz arcs in Figure 11-6. The crystal spins about the indicated symmetry axis, which remains horizontal.

Yet doubts persist. Do we really know what Lowitz arcs look like? Drawings are fallible, and Lowitz's drawing indeed contains obvious flaws — the intersecting circles instead of the circumscribed and 22° halos, for instance. And in spite of subsequent reports of Lowitz arcs [e.g., Ling, 1922], there seem to be no photographs of them. So one wonders exactly what Lowitz and others saw. If we accept the spinning crystal hypothesis, we introduce a theoretical complication — an exotic class of crystal orientations — based on questionable observations.

The simulation in Figure 11-7, made with Parry oriented columns instead of spinning plates,[4] offers an alternative explanation for Lowitz arcs: Lowitz arcs are just Parry arcs. This explanation requires no new crystal orientations and no new halos. Moreover, the Lowitz arcs (Parry arcs) would be natural in conjunction with the strong column crystal halos of the St. Petersburg display. True, the simulation is only a mediocre copy of Lowitz's drawing, and some excuses for the drawing need to be made. But I can believe, for example, that the strong Parry arc just below the circumscribed halo in the simulation would have been lost in the glare of the circumscribed halo, and I can believe, though less easily, that Lowitz missed the upper Parry arc.

---

4 Parry orientations are a subset of spinning plate orientations, so the similarities between the two simulations are not so surprising.

Display 6‑8 contains Parry arcs, admittedly faint, that remind me of Lowitz's drawing. Several displays — Display 11‑4, for example — suggest that Parry arcs can be much brighter and more colorful, more like what Lowitz described. None of this, of course, rules out spinning plates as an explanation of Lowitz arcs. But Parry arcs at high sun do occur, and they do look something like what Lowitz drew. In the absence of halo photographs resembling the spinning plate simulation, one has to wonder whether Lowitz didn't simply see Parry arcs.

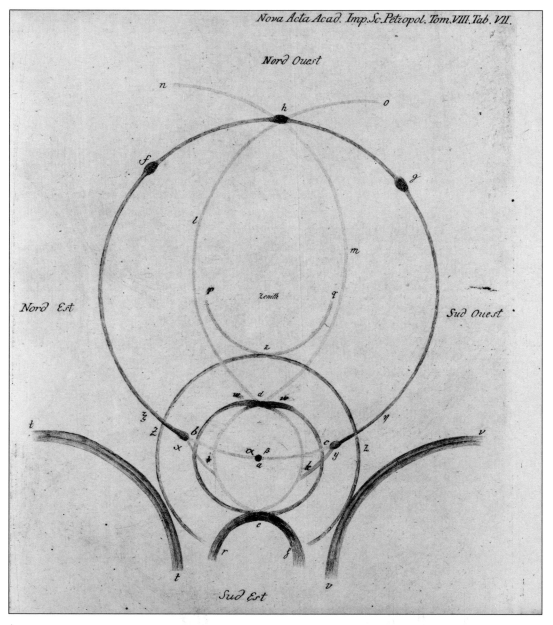

**Figure 11-5.** Lowitz's drawing of the St. Petersburg Display. The drawing is a composite made over several hours, and not all the halos shown were present at once. Lowitz mentioned, for example, that the circumscribed halo was absent early in the morning, he described the evolution of the upper tangent arc *wdw* with time, and he recorded the disappearance of the infralateral arcs *tt* and *vv*. The Lowitz arcs are *xi* and *yk*.
(Reproduced from Lowitz [1794].)

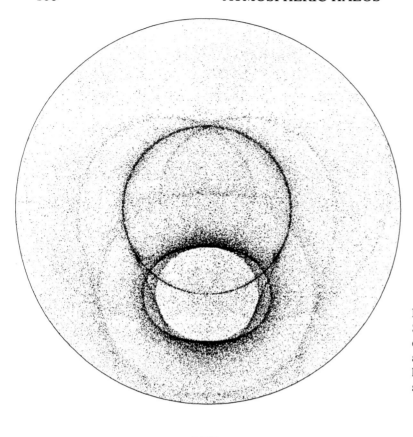

**Figure 11-6.** An attempt to simulate the St. Petersburg Display. Spinning plates, oriented plates, singly oriented columns, and randomly oriented crystals made the halos. The Lowitz arcs resulted from the spinning plates.

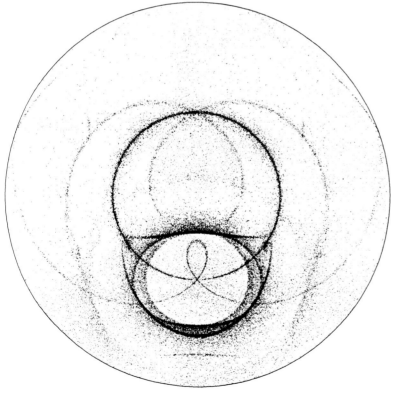

**Figure 11-7.** Another simulation of the St. Petersburg Display. The crystals are the same as in Figure 11-6 except that the spinning plates have been replaced by Parry oriented columns. The "Lowitz arcs" resulted from the Parry oriented columns — they are Parry arcs.

## Display 11-3 — The Saskatoon Display
### Saskatoon, Saskatchewan, December 3, 1970

This display was striking for the definition and intensity of its halos. It was described in two articles in the British magazine *Weather*, from which Figure 11-8 is taken [Ripley and Saugier, 1971; Evans and Tricker, 1972]. Several intriguing halos are depicted there, among them the Kern arc and the 46° parhelia.

**Kern arc.** The Saskatoon observation is one of only a few reports of the Kern arc, none substantiated by a photograph [e.g., Ling, 1922; Zamorsky, 1959]. Together with the circumzenith arc, the Kern arc is said to form a circle centered at the zenith. Simulations show that a Kern arc might arise in thick oriented plate crystals, but enormous numbers of crystals would be required, and the arc would be weak. The arc strengthens dramatically if the simulations use triangular plate crystals instead of hexagonal plates (Figures 11-9 and 11-10). Simulations made by Tränkle and Greenler [1987] show that a Kern arc might also arise through multiple scattering from oriented hexagonal plates.

**46° parhelia.** These same multiple scattering simulations also predict spot halos on the parhelic circle roughly 45° from the sun. In the simulations these new halos are parhelia of the parhelia. That is, an ordinary parhelion acts as a light source and produces two secondary parhelia about 22° away. One of them, the "46° parhelion", is located further out on the parhelic circle, and the other is at the sun. The 46° parhelia are rare, but their existence is documented with a photograph in the Evans and Tricker article.

Why are the Kern arc and the 46° parhelia included among the "mysteries" of the present chapter? For the Kern arc, the absence of photographs is disconcerting. And for the 46° parhelia, their occasional reported brightness seems inconsistent with their interpretation as secondary halos. Display 11-4 is an example.

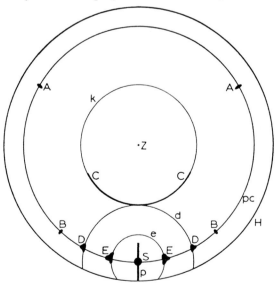

**Figure 11-8.** The Saskatoon Display. The Kern arc is *k*, and the 46° parhelia are *DD*. (Reproduced from Ripley and Saugier [1971], with permission of *Weather* and the Royal Meteorological Society)

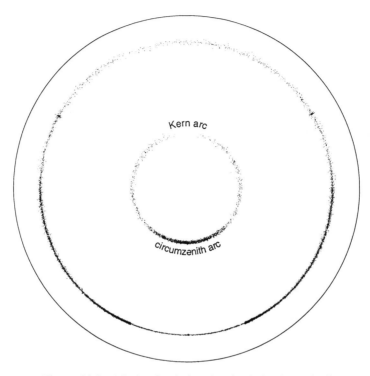

**Figure 11-9.** All-sky simulation showing halos theoretically expected from oriented triangular plate crystals.

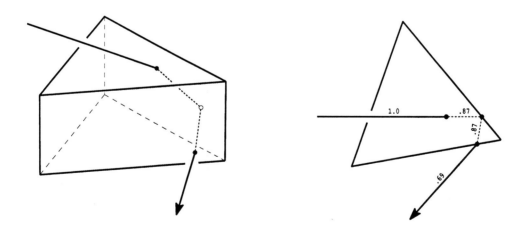

**Figure 11-10.** (Left) Triangular plate crystal and ray path for the Kern arc. Like a ray for the circumzenith arc, the ray enters the top basal face and exits a prism face, but here there is an intervening internal reflection from a prism face. (Right) Same ray path seen from above. The internal reflection is total and the outgoing ray is strong.

## Display 11-4
## Antarctica, November 29, 1958

This spectacular display is one of several seen by J. R. Blake during an Antarctic expedition in 1958-1959. Blake's drawing of the display is reproduced below. These notes are also his:

> The 22° and 46° halos were both very brightly colored, and both 22° and 46° parhelia were present, both pairs being very bright. The vertical pillar extended from the horizon to the top of the 22° halo, which was brilliant. From this point extended the upper contact arc which merged smoothly with Parry's arc; both these arcs were colored and very bright... The circumzenithal arc was also present, being very brightly colored.

> A fairly bright, colored arc, passing through or close to the zenith and concave to the sun was visible, the colors being very pure and distinct...

> The full 180° of the parhelic circle was visible on either side of the sun, being intersected at approximately 90° from the sun by a pair of white pillars extending from the horizon to slightly above the parhelic ring. The points of intersection resembled mock-suns, though white.

> Also at 180° there existed a somewhat fainter, white pillar extending to just above the horizon, the point of intersection with the parhelic circle again being brighter.

Since secondary halos should be weak, the interpretation of 46° parhelia as secondary halos, offered in connection with the Saskatoon Display, seems inadequate for the bright 46° parhelia described by Blake. Also puzzling are the two pillar-like halos at 90° from the sun (Hevel's halo?!). But the strangest halo is the colored arc near the zenith. I am afraid I have no explanations for these very rare halos.[5] Next time perhaps someone will get a halo photograph and a crystal sample.

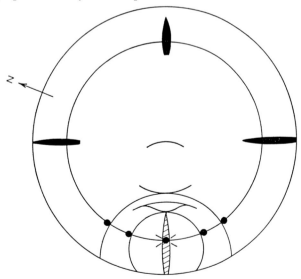

**Figure 11-11.** Halo display drawn by J. R. Blake. The halo near the zenith and the pillar-like halos at right and left are unexplained. (Reproduced from Blake [1961], with permission of the Australian Antarctic Division)

---

5 Horizontal column crystals with two prism faces *vertical* would produce 46° parhelia, but they would produce other halos as well — halos that have never been reported. See Figure C-32 of Appendix C.

**An invitation.** Puzzling halo reports occur sporadically throughout the scientific literature. Most of the "halos" were reported only once or twice, and hardly any were photographed. Some were probably wispy clouds, some were products of active imaginations (people *want* to see rare halos), some were inaccurately recorded, and some perhaps were hoaxes by the "observer." Nevertheless, the experience of the past two decades, during which many rare and previously questionable halos were photographed for the first time, suggests that among the old observations are correct accounts of real halos, waiting to be substantiated by photographs of similar phenomena. I hope that some of you will be the ones to do so.

# APPENDIX A

## SUGGESTIONS FOR PHOTOGRAPHING HALOS

Anyone who sees a rare halo can obtain useful photographs. There are few tricks for good halo photography; the crucial element is the display itself.

I recommend slow-speed color slide film. Be sure to block the sun, in spite of the negative aesthetic impact. Take one exposure as recommended by your light meter and a second underexposed one f-stop. Additional exposures are necessary for fisheye photographs, due to the wide variation in sky brightness. Do not stint on film, since most displays vary during their course. Note the time and place of each photograph, so that the sun elevation can be computed later.

Your first attempt at halo photography will reveal the immensity of halos and the consequent need for wide angle lenses. A 28-mm lens has a wide enough field of view to include both parhelia. A 21-mm lens will capture the entire 22° halo, and a 15-mm lens, the supralateral arc. When buying lenses, you may have to compromise for budget's sake, but zoom lenses are not so good if angular measurements are ever to be made from the photographs. I suggest a 28-mm lens for a first purchase, followed by a 15 mm. High-quality lenses are not necessary.

Finally, two obvious but crucial reminders: Watch the sky and keep the camera with you. Good luck!

106

## APPENDIX B

## SOME HISTORY

As explained in Chapter 1 and elsewhere, the computer can be programmed to look for crystal orientations and ray paths responsible for a given halo. That is the approach used in this book. For many halos, however, the orientations and ray paths have long been known, thanks to the ingenuity of many scientists over the past several centuries. Appendix B contains some halo history and, in particular, traces the discovery of crystal orientations and ray paths for the various halos.

E. Mariotte [1717] in the late seventeenth century was the first to attribute halos to prismatic ice crystals. He showed how randomly oriented crystals would cause the 22° halo and how crystals with axes vertical would cause the parhelia.

By the early nineteenth century Thomas Young [1807] knew that crystals with horizontal axes would cause the upper tangent arc. He also knew that reflections from vertical crystal faces would cause the parhelic circle. Young credits Cavendish for the correct ray path for the 46° halo.

The mid-nineteenth century saw the publication of M. A. Bravais' classic *Mémoire sur les Halos...*. In it, Bravais [1847] described crystal orientations and ray paths for many common halos, and he calculated halo shapes as a function of sun elevation. He used pyramidal crystals to explain odd radius halos, and he discussed secondary halos. The memoir also summarizes much of the early history of halo studies, both theoretical and observational.

The early twentieth century was marked by the work of C. S. Hastings and, especially, Alfred Wegener. Among other achievements, they share credit for explaining the (upper suncave) Parry arc. The arc had gone unexplained for almost a century after the sighting of one by W. E. Parry [1821] in the arctic. Then Hastings [1902, 1920] suggested that the Parry arc might arise in crystals with two prism faces horizontal, that is, in Parry oriented crystals. W. Brand and Wegener [1912], probably independently of Hastings, used Parry orientations to analyze a Parry arc that Wegener had observed in Greenland in 1908. They calculated its theoretical angular elevation and found good agreement with the observed halo.

Hastings is also credited with the explanation of the subsequently named Hastings arc. However, the anthelic arcs that he was trying to explain were probably Wegener arcs rather than Hastings arcs. That is, they were more likely due to singly oriented columns than to Parry oriented columns. In general, Hastings overrated Parry oriented columns at the expense of singly oriented columns. For example, he mistakenly attributed the infralateral arcs to Parry oriented columns.

Hastings knew the common ray path for the 120° parhelia, and he used spinning crystals to explain Lowitz arcs, though he credits the idea to Galle.

In 1925 Wegener published his elegant *Theorie der Haupthalos.* In it he presented a general setting for studying halos. He showed how to calculate the shapes of most of the known halos, illustrating his results with detailed plots of the tangent arcs, infralateral arcs, Parry arc, and Wegener arc. The Wegener arc, of course, was being explained for the first time. Wegener also explained the subparhelia, and he foresaw the possibility of several new halos: the subparhelic circle, the sunvex Parry arc, the Parry infralateral arcs, and the subhelic arc. For the subhelic arc, he even calculated its shape, but he doubted it would be bright enough to be seen. Ironically, he had probably seen one in the Greenland display mentioned above, without recognizing it.

W. J. Humphreys [1940] is apparently responsible for explaining the heliac arc.

P. Putnins [1934] published the theory of the upper sunvex and lower Parry arcs.

R. A. R. Tricker [1973] developed the theory of the antisolar arc and of the anthelic arc now named after him. The antisolar arc had been seen by J. R. Blake [1961] in Antarctica.

Using a computer, Robert Greenler and E. Tränkle [1984] simulated the diffuse arcs and found their ray paths. Blake [1961] and G. H. Liljequist [1956] had seen the arcs previously.

Rudimentary Parry supralateral arcs — spots on the supralateral arc — had been seen by Parry himself. Hastings mentioned Parry's observation in his book [Hastings, 1902] and correctly attributed the Parry supralateral arcs to Parry oriented crystals. But perhaps because Hastings incorrectly analyzed the (non-Parry) infralateral arcs, or because he failed to discuss the Parry supralateral arcs in his later journal article [Hastings, 1920], which is more widely known than his book, both Parry's observation and Hastings' explanation escaped notice of later writers. Thus the sightings of Parry supralateral arcs in Chapter 3 came as a surprise.

## APPENDIX C

## HALO SIMULATIONS AT SELECTED SUN ELEVATIONS

This appendix illustrates the theoretical dependence of halos on sun elevation. While theory should not dictate what you see, the simulations can suggest what to look for in skywatching. In the simulations, however, the crystals were well-formed prisms, and the orientations of the column crystals were nearly perfect. The simulations therefore show the most that can be theoretically expected from the given crystals. For reasons given in Chapter 5, real halo displays usually fall short of these optimal predictions.

**Figure C-1 through Figure C-10:** Halos from oriented plate crystals.
The mean c/a ratio of the crystals is 0.2. Except for Figure C-10, the crystal tilts are 2°.
(For a prismatic crystal, the value c is is defined to be the distance between the basal faces, and the value a is the distance across a basal face. An equidimensional crystal therefore has c/a about equal to 1, whereas a plate crystal has c/a considerably less than 1, and a column, considerably greater. This geometrical c/a ratio should not be confused with the crystallographic c/a ratio mentioned in Chapter 10.)

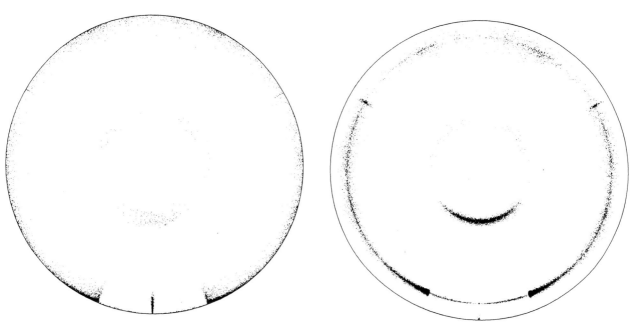

**Figure C-1.** Sun elevation 0°.          **C-2.** Sun elevation 10°.

(oriented plates continued)

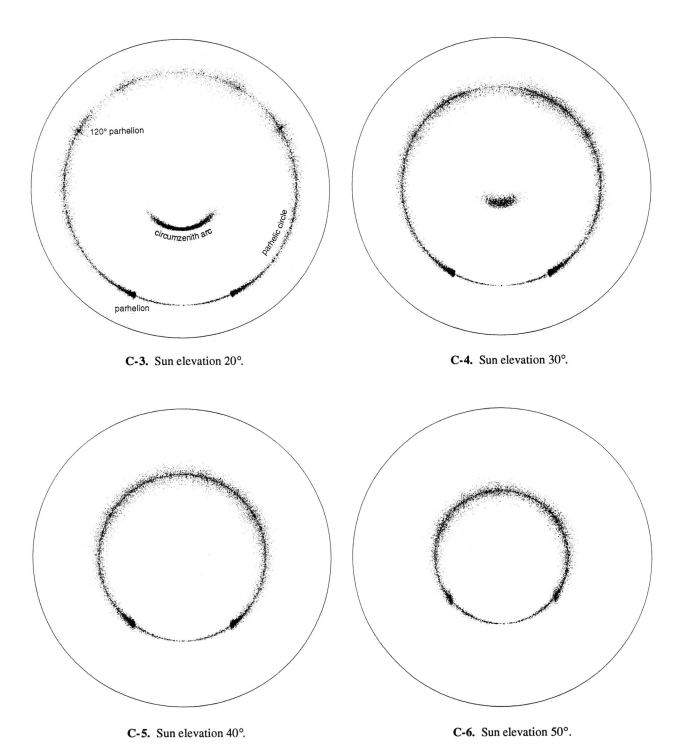

**C-3.** Sun elevation 20°.

**C-4.** Sun elevation 30°.

**C-5.** Sun elevation 40°.

**C-6.** Sun elevation 50°.

(oriented plates continued)

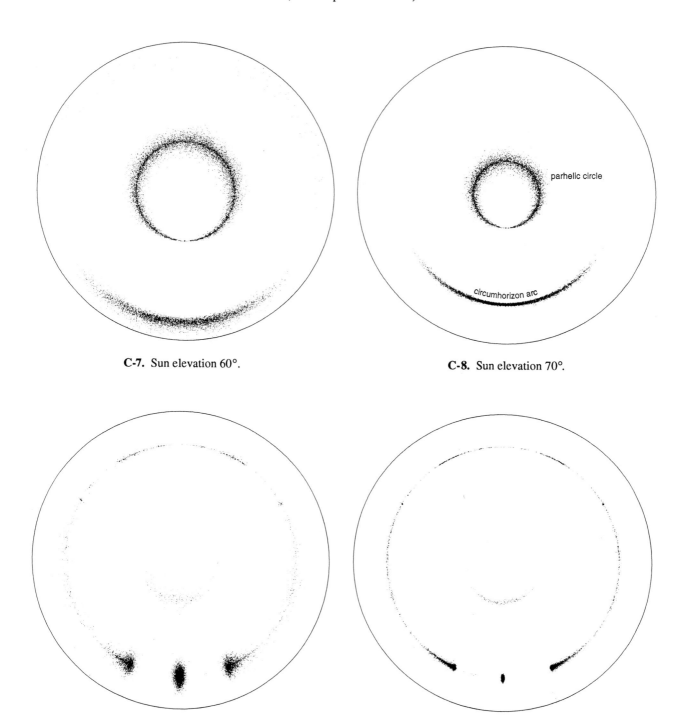

**C-7.** Sun elevation 60°.

**C-8.** Sun elevation 70°.

**C-9.** Subhorizon halos. Sun elevation 20°. Compare with Figure C-3.

**C-10.** Same as Figure C-9 but with crystal tilts of 0.5° instead of 2°.

**Figure C-11 through Figure C-22:** Halos from singly oriented columns.
The mean c/a ratio of the crystals is 2.0. Except for Figure C-20, the crystal tilts are 0.1°.

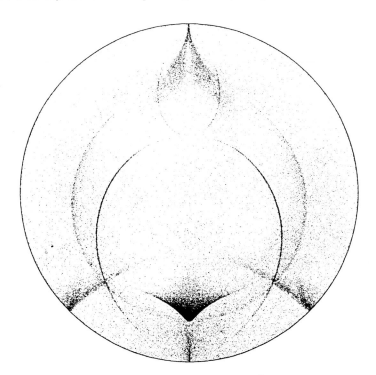

**Figure C-11.**  Sun elevation 0°.

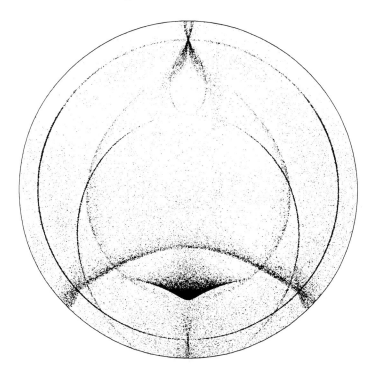

**C-12.**  Sun elevation 10°.

(singly oriented columns continued)

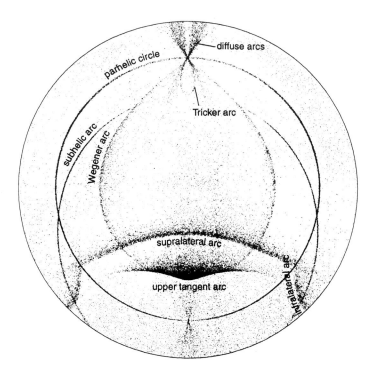

**C-13.** Sun elevation 20°.

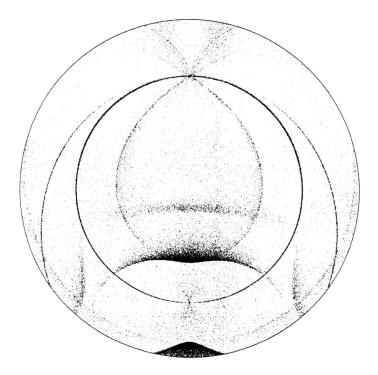

**C-14.** Sun elevation 30°.

(singly oriented columns continued)

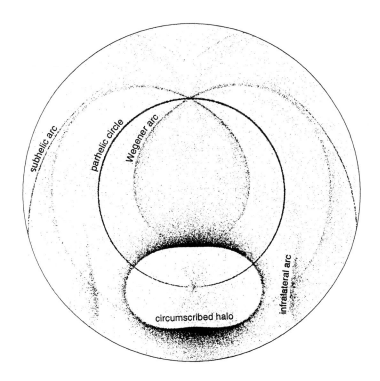

**C-15.** Sun elevation 40°.

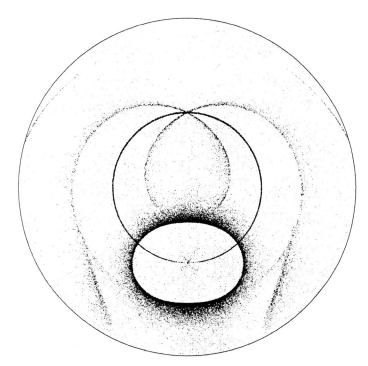

**C-16.** Sun elevation 50°.

(singly oriented columns continued)

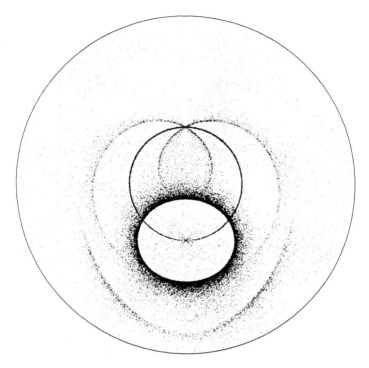

**C-17.** Sun elevation 60°.

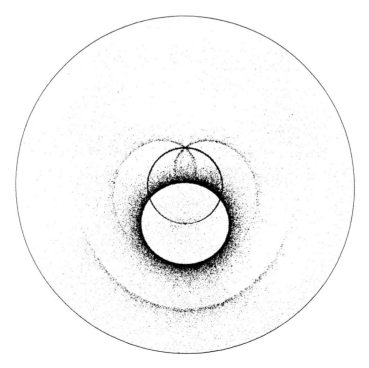

**C-18.** Sun elevation 70°.

(singly oriented columns continued)

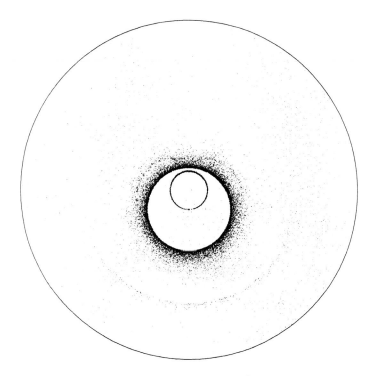

**C-19.**  Sun elevation 80°.

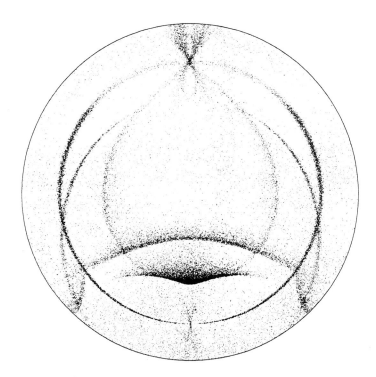

**C-20.**  Same as Figure C-13 but with crystal tilts
of 0.5° instead of 0.1°.

(singly oriented columns continued)

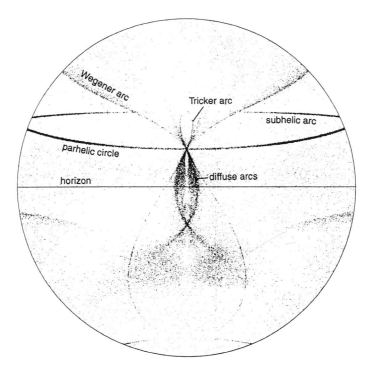

**C-21.** Fisheye view centered on the point on the horizon opposite the sun. Subhorizon halos are included. The sun elevation is 20°.

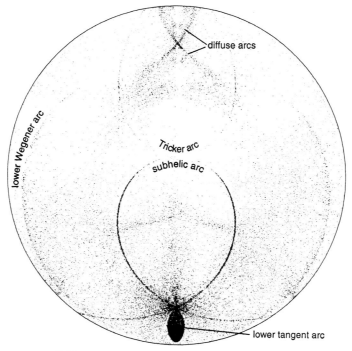

**C-22.** Subhorizon halos. Sun elevation 20°. The lower tangent arc is upside down in this view.

**Figure C-23 through Figure C-31:** Halos from Parry oriented columns.
The mean c/a ratio of the crystals is 2.0, their tilts are 0.1°, and their rotation angles are 0.1°.
These simulations are probably unrealistically strong; many of the faint halos have never been reported.

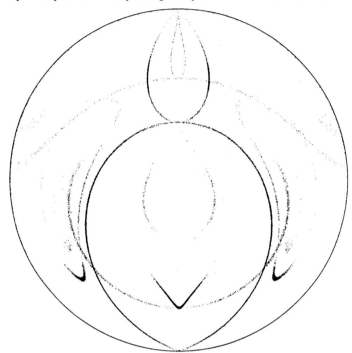

**Figure C-23.**  Sun elevation 0°.

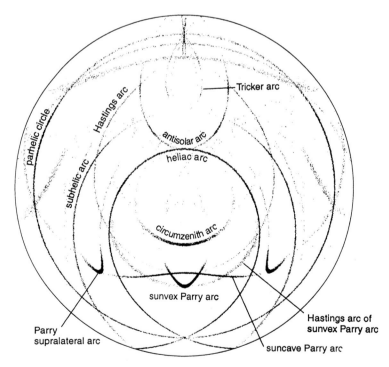

**C-24.**  Sun elevation 10°.

(Parry oriented columns continued)

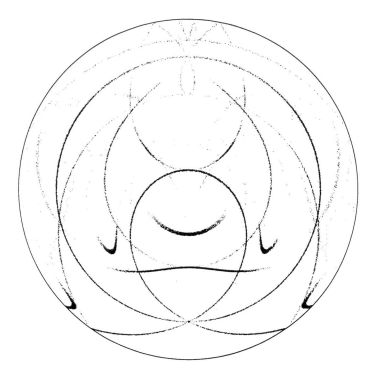

**C-25.** Sun elevation 20°.

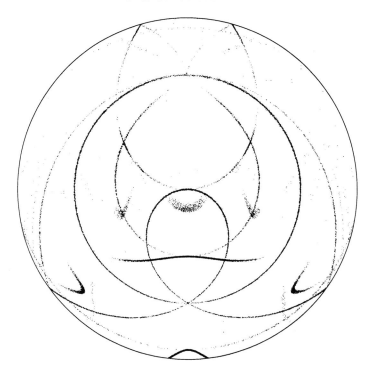

**C-26.** Sun elevation 30°.

(Parry oriented columns continued)

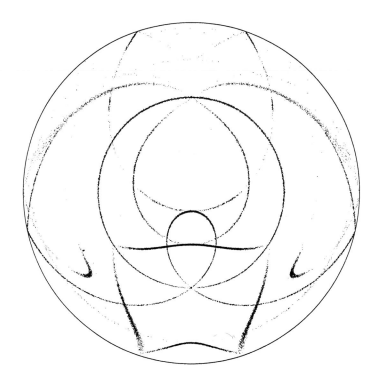

**C-27.** Sun elevation 40°

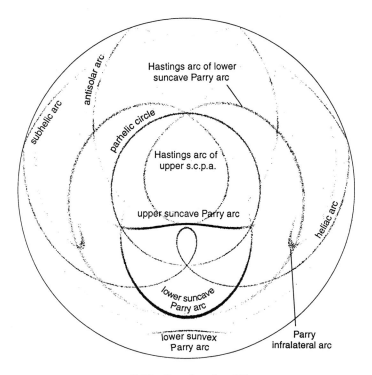

**C-28.** Sun elevation 50°.

(Parry oriented columns continued)

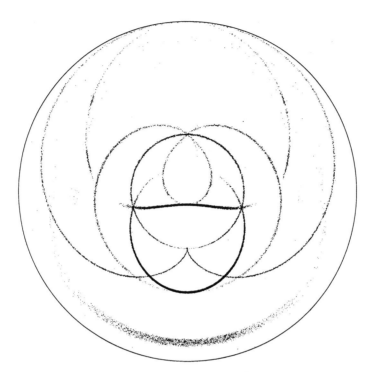

**C-29.** Sun elevation 60°

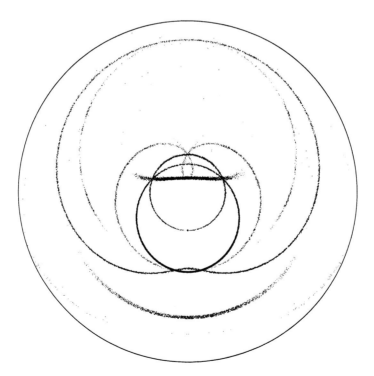

**C-30.** Sun elevation 70°.

(Parry oriented columns continued)

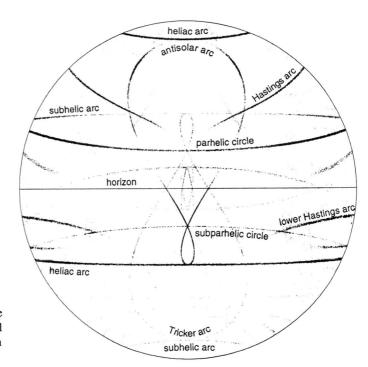

**C-31.** Fisheye view centered on the point on the horizon opposite the sun. Notice the beautiful symmetries between the halos above the horizon and those below. Sun elevation 20°.

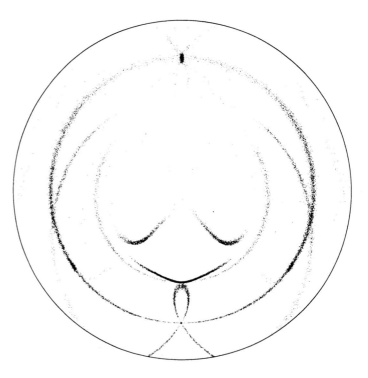

**Figure C-32.** Halos theoretically expected from horizontal columns with two prism faces vertical. These "alternate Parry" orientations provide an attractive explanation for 46° parhelia and the anthelion, but there seems to be no independent evidence for the existence of such orientations. Furthermore, neither the strong "alternate Parry arc" nor the "alternate Parry supralateral arcs" predicted here have been reported. However, an alternate Parry arc might be masked by a strong upper tangent arc, and alternate Parry supralateral arcs might be too weak to be seen. The existence of alternate Parry orientations should therefore be entertained as an outside possibility. The tilts and rotation angles of the crystals in the simulation are both 0.5°, more than in the preceding simulations. The sun elevation is 20°.

## APPENDIX D

## ICE CRYSTAL HABIT

There are many possible types of ice crystals. Laboratory experiments show that crystal type depends on the temperature and humidity under which the crystal forms and grows. Some of this dependence of crystal type on atmospheric conditions is summarized in the diagram below. The diagram is relevant to halo studies because, as explained in Chapter 5, not all crystal types are good halo makers; most, in fact, are not. The diagram helps to explain why good halo displays are not common.

Having looked at ice crystals during many halo displays, I have some reservations about the complete applicability of the diagram to the world outside the laboratory. But the details are not crucial here; what matters is that crystal type does depend in some way on conditions, and that without the proper conditions, one gets poor halos.

One other point in connection with crystal formation: Atmospheric water droplets normally remain liquid at temperatures well below the "freezing point" of water. The diagram does not imply that a crystal will indeed form from a water droplet under the given conditions. So although a cloud may be cold, it need not consist of ice crystals. Again, this contributes to the scarcity of halos.

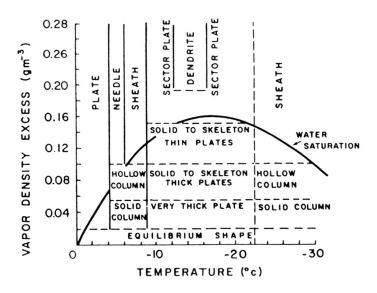

Variation of ice crystal type with temperature and humidity.
(adapted from Pruppacher and Klett [1978, p. 30])

## APPENDIX E

## RAY PATHS FOR HALOS

Every plotted dot in a halo simulation is determined by a ray path through a crystal. The computer can be programmed to sort and count the ray paths and to identify the halo associated with each. It thus reveals the distribution of ray paths making the halos in the simulation.

Each of the first six tables in this appendix gives the distribution of ray paths for a particular simulation. Thus Table 1 is for an all-sky simulation made with oriented plate crystals. It shows, for instance, that ray paths for the circumzenith arc are all of one type — that shown in Figure 1-8. On the other hand, it reveals a variety of ray paths, some quite complicated, for the parhelic circle. (The tables are explained in detail below.)

A ray path is described by the sequence of crystal faces that it encounters. Crystal faces are numbered as shown; basal faces are *1* and *2*, and prism faces are *3* through *8*. For example, the Tricker arc path at the top in Figure 2-25 is *35714* (or *46815*, or *57316*, etc.).

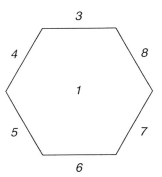

Some simplification and grouping of ray paths is necessary. In the tables, ray paths are grouped into subtypes and types. The guiding principle behind the grouping is that paths should be considered similar if their outgoing rays have the same direction. Such paths are therefore grouped together.

To obtain the subtype of a ray path, start with the ray path — a sequence of face numbers — and delete pairs of internal reflections that involve opposite basal faces. Write any remaining internal basal face reflection in the second position of the sequence. Then delete pairs of successive reflections that involve opposite prism faces. Except for Parry oriented crystals and spinning plates, renumber the prism faces so that face *3* is the first prism face encountered by the ray. Examples: The parhelion ray path *4218* is subtype *37*. The Tricker arc path *1372531* is subtype *1237531*. The column parhelic circle path *36236* is subtype *326*.

Then to obtain the type, delete any remaining sequences of reflections that do not change the direction of the ray. Examples: For the paths of the preceding paragraph, the types are the same as the subtypes. For the antisolar arc path *4287654386*, the type is *4286*, since the reflections *876543* leave the direction of the ray unchanged, each pair *87*, *65*, and *43* deviating the ray by 120° as seen from the end of the crystal (Figure 3-13, bottom right). For additional examples see Table 7, which gives the path type for each ray path figure.

Still shorter sequences appear parenthetically in the tables. The plate crystal parhelic circle path *387* is "like *3*" (Table 1); this means that the direction of the outgoing ray is as if the incoming ray had simply reflected externally from crystal face (*5*, which is then renumbered) *3*. A bar over a numeral indicates an idealized basal face reflection; the antisolar arc path *3287* is "like $\bar{5}$" (Table 6 and Figure 3-13, top), because the direction of the outgoing ray is as if the incoming ray had reflected externally first from face *5* and then from a basal face — a physical impossibility but a useful convention.

To read the tables, consider Table 1, which gives the distribution of ray paths through oriented plates. Of the four halos in the table, the circumzenith arc is simplest. Its 121 paths are all type and subtype *13*. The paths themselves could be *13*, or *14*, or *15*, etc. Figure 1-8 shows such a path.

The parhelion is nearly as simple; its 118 paths are type and subtype *37*. However, the table discriminates within subtypes by indicating deleted reflections as suffixes. The 92 paths in entry *37-2b* thus have two extra basal face reflections. The paths could be *3217*, or *4218*, etc.

The parhelic circle is surprisingly complex, with nine path types and even more subtypes. Most of its paths, however, are "like *3*" — roughly equivalent to an external reflection from a prism face.

The 120° parhelion is simple, its paths all being like the one in Figure 1-18.

The successive shortenings — from ray path to subtype, to type, to "like" — do not change the direction of the outgoing ray, but they must be interpreted with care. For example, the ray path *35673* is "like *3*", in the sense that paths *35673* and *3* have the same direction for their outgoing rays, so long as the incoming rays are the same. However, the intensities of the two outgoing rays are not the same. Also, not every incoming ray that impinges on face *3* can traverse path *35673*. So the "halo" due to type *3* rays differs from that due to *35673*. In fact, if the crystals are oriented plates, then type *3* rays distribute light over the entire parhelic circle, while *35673* rays make a bright parhelic circle segment about 30° to the left of the anthelic point (Figure C-3).

South is taken to be in the direction of the sun. With this understanding, ray paths lighting the eastern half of the sky are excluded from the tables. Singly oriented columns and Parry oriented columns have face *1* visible from the east, Parry oriented columns have face *3* on top, and oriented plates have face *1* on top. These conventions account for asymmetries in the tables. In Table 1, for example, the left parhelion path *35* is absent.

The sun elevation for the tables is 20°. For plate crystals the mean c/a ratio is 0.2, and for column crystals, 2.0. Changing these parameters can change the tables.

**Table 1.** Ray paths in oriented plate crystals.

| Number of Paths | | Type or Subtype | | Halo | Figure |
|---|---|---|---|---|---|
| 118 | | *3 7* | | parhelion | |
| | 26 | *3 7* | | | |
| | 92 | *3 7* - 2b | | | |
| 121 | | *1 3* | | circumzenith arc | Figure 1-8 |
| 47 | | *3* | | parhelic circle | Figure 1-17 top |
| 150 | | *1 3 2* | ( like *3* ) | parhelic circle | |
| | 140 | *1 3 2* | | | Figure 1-17 bottom |
| | 9 | *1 3 2* - 2b | | | |
| | 1 | *1 3 2* - 4b | | | |
| 1 | | *3 6 3* | ( like *3* ) | parhelic circle | |
| | 1 | *3 6 3* - 4b | | | |
| 9 | | *3 7 5* | ( like *3* ) | parhelic circle | |
| | 6 | *3 7 5* - 2b | | | |
| | 2 | *3 7 5* - 4b | | | |
| | 1 | *3 7 6 5 4 3 8 7 5* - 6b | | | |
| 9 | | *3 8 7* | ( like *3* ) | parhelic circle | |
| | 2 | *3 8 7* | | | |
| | 7 | *3 8 7* - 2b | | | |
| 11 | | *3 5 6 7 3* | ( like *3* ) | parhelic circle | |
| | 4 | *3 5 6 7 3* - 2b | | | |
| | 6 | *3 5 6 7 3* - 4b | | | |
| | 1 | *3 5 6 7 3* - 6b | | | |
| 13 | | *3 5 6 7 8 3 5* | ( like *3* ) | parhelic circle | |
| | 7 | *3 5 6 7 8 3 5* - 4b | | | |
| | 2 | *3 5 6 7 8 3 5* - 6b | | | |
| | 2 | *3 5 6 7 8 3 5* - 8b | | | |
| | 1 | *3 5 6 7 8 3 4 5 6 7 8 3 5* - 6b | | | |
| | 1 | *3 5 6 7 8 3 4 5 6 7 8 3 5* - 12b | | | |
| 6 | | *3 7 6 5* | | parhelic circle with net refraction | |
| | 4 | *3 7 6 5* - 2b | | | |
| | 2 | *3 7 6 5* - 4b | | | |
| 6 | | *3 8 7 5* | | parhelic circle with net refraction | |
| | 6 | *3 8 7 5* - 2b | | | |
| 9 | | *1 3 8 2* | | 120° parhelion | Figure 1-18 |
| 500 total | | | | | |

Types are shown in bold. Refer to Figure C-3 of Appendix C for the halos. A ray path type for a halo need not spread light equally over the halo. For example, ray paths of types *35673*, *3765*, and *3875* make the bright segment on the parhelic circle about 30° to the left of the anthelic point, rather than lighting the entire parhelic circle.

## Table 2. Ray paths in oriented plate crystals; subhorizon halos.

| Number of Paths | | Type or Subtype | | Halo | Figure |
|---|---|---|---|---|---|
| 190 | | *1* | | subsun | Figure 7-3 top |
| 68 | | *1 2 1* | ( like *1* ) | subsun | |
| | 65 | *1 2 1* | | | |
| | 3 | *1 2 1* - 2b | | | |
| 247 | | *3 2 6* | ( like *1* ) | subsun | |
| | 229 | *3 2 6* | | | Figure 7-3 bottom |
| | 18 | *3 2 6* - 2b | | | |
| 366 | | *3 2 7* | | subparhelion | |
| | 355 | *3 2 7* | | | |
| | 11 | *3 2 7* - 2b | | | |
| 14 | | *1 2 3* | | subcircumzenith arc | |
| 32 | | *1 2 3 1* | ( like $\bar{3}$ ) | subparhelic circle | |
| | 31 | *1 2 3 1* | | | Figure 7-7 top |
| | 1 | *1 2 3 1* - 2b | | | |
| 7 | | *3 2 7 5* | ( like $\bar{3}$ ) | subparhelic circle | |
| | 2 | *3 2 7 5* | | | |
| | 5 | *3 2 7 5* - 2b | | | |
| 40 | | *3 2 8 7* | ( like $\bar{3}$ ) | subparhelic circle | |
| 20 | | *3 2 5 6 7 3* | ( like $\bar{3}$ ) | subparhelic circle | |
| | 15 | *3 2 5 6 7 3* - 2b | | | Figure 7-7 bottom |
| | 3 | *3 2 5 6 7 3* - 4b | | | |
| | 1 | *3 2 5 6 7 8 3 4 5 6 7 3* - 6b | | | |
| | 1 | *3 2 5 6 7 8 3 4 5 6 7 8 3 4 5 6 7 3* - 10b | | | |
| 1 | | *3 2 5 7 8 3* | ( like $\bar{3}$ ) | subparhelic circle | |
| | 1 | *3 2 5 7 8 3* - 4b | | | |
| 5 | | *3 2 5 6 7 8 3 5* | ( like $\bar{3}$ ) | subparhelic circle | |
| | 2 | *3 2 5 6 7 8 3 5* - 2b | | | |
| | 2 | *3 2 5 6 7 8 3 5* - 4b | | | |
| | 1 | *3 2 5 6 7 8 3 4 5 6 7 8 3 4 5 6 7 8 3 5* - 10b | | | |
| 2 | | *3 2 7 6 5* | | subparhelic circle with net refraction | |
| | 1 | *3 2 7 6 5* | | | |
| | 1 | *3 2 7 6 5* - 2b | | | |
| 5 | | *3 2 8 7 5* | | subparhelic circle with net refraction | |
| | 3 | *3 2 8 7 5* | | | |
| | 2 | *3 2 8 7 5* - 2b | | | |
| 2 | | *1 2 3 8 1* | | sub 120° parhelion | |
| 1 | | *3 2 7 6 4* | | sub 120° parhelion | |
| | 1 | *3 2 7 6 5 4 3 7 4* - 10b | | | |
| 1000 total | | | | | |

Refer to Figure C-10 for the halos, which are obvious analogues of familiar above-horizon halos. Ray paths of types *325673*, *32765*, and *32875* make a bright segment on the subparhelic circle, like the one in Figure 7- 6 but on the left side of the antisolar point.

**Table 3.** Ray paths in singly oriented columns.

| Number of Paths | | Type or Subtype | | Halo | Figure |
|---|---|---|---|---|---|
| 276 | | *3 7* | | upper tangent arc | Figure 2-7 |
| 89 | | *3 2* | | supralateral arc | |
| 25 | | *3 1* | | supralateral arc | Figure 2-13 |
| 32 | | *2 3* | | infralateral arc | |
| | 31 | 2 3 | | | |
| | 1 | 2 3 - 2p | | | |
| 24 | | *2 3 5 1* | | subhelic arc | Figure 2-23 |
| 28 | | *2 3 8 1* | | subhelic arc | |
| 41 | | *3 2 7* | | Wegener arc | Figure 2-21 |
| 44 | | *1* | | parhelic circle | Figure 2-15 left |
| 139 | | *3 2 6* | ( like *1* ) | parhelic circle | |
| | 137 | 3 2 6 | | | Figure 2-15 right |
| | 1 | 3 2 7 6 5 3 6 | | | |
| | 1 | 3 2 5 6 7 3 6 | | | |
| 12 | | *3 1 7 5* | ( like $\bar{3}$ ) | diffuse B | Figure 2-26 bottom |
| 14 | | *3 2 7 5* | ( like $\bar{3}$ ) | diffuse B | |
| | 8 | 3 2 7 5 | | | |
| | 6 | 3 2 7 6 5 4 3 8 7 5 | | | |
| 21 | | *3 1 5 6 7 3* | ( like $\bar{3}$ ) | diffuse A | Figure 2-26 top |
| 7 | | *3 2 5 6 7 8 3 5* | ( like $\bar{3}$ ) | | |
| 83 | | *3* | | | |
| 57 | | *2 3 1* | ( like *3* ) | | |
| | 55 | 2 3 1 | | | |
| | 2 | 2 3 1 - 2p | | | |
| 13 | | *3 7 5* | ( like *3* ) | | |
| | 11 | 3 7 5 | | | |
| | 2 | 3 7 6 5 4 3 8 7 5 | | | |
| 24 | | *3 8 7* | ( like *3* ) | | |
| 8 | | *3 5 6 7 3* | ( like *3* ) | | |
| | 6 | 3 5 6 7 3 | | | |
| | 2 | 3 5 6 7 3 - 2b | | | |
| 5 | | *2 3 8 6* | | | |

58 paths in 34 remaining types

1000 total

Refer to Figure C-13 for the halos.

**Table 4.** Ray paths in singly oriented columns that light the 60° wide by 40° high region of sky centered at the anthelic point.

| Number of Paths | | Type or Subtype | | Halo | Figure |
|---|---|---|---|---|---|
| 36 | | *3 2 7* | | Wegener arc | Figure 2-21 |
| 69 | | *1* | | parhelic circle | Figure 2-15 left |
| 65 | | *3 2 6* | ( like *1* ) | parhelic circle | Figure 2-15 right |
| 126 | | *3 1 7 5* | ( like $\bar{3}$ ) | diffuse B | |
| | 123 | *3 1 7 5* | | | Figure 2-26 bottom |
| | 2 | *3 1 7 6 5 3 7 5* | | | |
| | 1 | *3 1 7 6 5 3 7 5* - 2b | | | |
| 63 | | *3 2 7 5* | ( like $\bar{3}$ ) | diffuse B | |
| | 61 | *3 2 7 5* | | | |
| | 1 | *3 2 7 6 5 4 3 8 7 5* - 2b | | | |
| | 1 | *3 2 7 6 5 4 3 8 7 6 5 4 3 8 7 5* - 2b | | | |
| 307 | | *3 1 5 6 7 3* | ( like $\bar{3}$ ) | diffuse A | |
| | 306 | *3 1 5 6 7 3* | | | Figure 2-26 top |
| | 1 | *3 1 5 6 7 3* - 2p | | | |
| 46 | | *3 2 5 6 7 3* | ( like $\bar{3}$ ) | diffuse A | |
| 18 | | *3 1 5 7 4* | | Tricker arc | Figure 2-25 top |
| 10 | | *3 1 6 8 4* | | Tricker arc | |
| 11 | | *2 1 3 5 2* | | Tricker arc | |
| 8 | | *1 2 3 7 5 3 1* | ( like *1 2 3 5 1* ) | Tricker arc | Figure 2-25 bottom |
| 21 | | *3* | | | |
| 16 | | *3 7 5* | ( like *3* ) | | |
| 39 | | *3 5 6 7 3* | ( like *3* ) | | |
| 25 | | *3 7 6 5* | | | |
| 27 | | *3 8 7 5* | | | |
| 11 | | *3 1 7 6 5* | | | |
| 14 | | *3 1 8 7 5* | | | |
| 88 paths in 29 remaining types | | | | | |
| 1000 total | | | | | |

**Table 5.** Ray paths in singly oriented columns that light the anthelion.

| Number of Paths | Type or Subtype | | Halo | Figure |
|---|---|---|---|---|
| 6 | *1* | | parhelic circle | Figure 2-15 left |
| 5 | *3 2 6* | ( like *1* ) | parhelic circle | Figure 2-15 right |
| 14 | *3 1 7 5* | ( like *$\bar{3}$* ) | diffuse B | Figure 2-26 bottom |
| 15 | *3 2 7 5* | ( like *$\bar{3}$* ) | diffuse B | |
| 22 | *3 1 5 6 7 3* | ( like *$\bar{3}$* ) | diffuse A | Figure 2-26 top |
| 20 | *3 2 5 6 7 3* | ( like *$\bar{3}$* ) | diffuse A | |
| 4 | *3 2 6 8 4* | | Tricker arc | |

14 paths in 11 remaining types

100 total

**Table 6.** Ray paths in Parry oriented columns.

| Number of Paths | | Type or Subtype | | Halo | Figure |
|---|---|---|---|---|---|
| 207 | | *3 7* | | Parry arc | Figure 3-8 |
| 68 | | *3 2* | | circumzenith arc | |
| 55 | | *4 2* | | Parry supralateral arc | |
| 61 | | *2 7* | | Parry infralateral arc | |
| 15 | | *4* | | heliac arc | |
| 30 | | *2 7 1* | ( like *4* ) | heliac arc | |
| 47 | | *5* | | heliac arc | Figure 3-12 top |
| 61 | | *3 8 7* | ( like *5* ) | heliac arc | Figure 3-12 bottom |
| 7 | | *2 7 3 4 1* | ( like *5* ) | heliac arc | |
| 7 | | *8* | | heliac arc | |
| 41 | | *2 8 7 1* | | subhelic arc | |
| 5 | | *2 6 7 8 3 1* | ( like *2 8 7 1* ) | subhelic arc | |
| 54 | | *3 2 7* | | Hastings arc | Figure 3-14 |
| 41 | | *1* | | parhelic circle | |
| 7 | | *3 2 6* | ( like *1* ) | parhelic circle | |
| | 6 | *3 2 6* | | | |
| | 1 | *3 2 6* - 2p | | | |
| 212 | | *4 2 7* | ( like *1* ) | parhelic circle | |
| | 208 | *4 2 7* | | | |
| | 1 | *4 2 7* - 2p | | | |
| | 1 | *4 2 7 3 4 5 7* | | | |
| | 1 | *4 2 7 5 4 3 7* | | | |
| | 1 | *4 2 6 7 8 4 7* | | | |
| 6 | | *3 2 8 7* | ( like $\overline{5}$ ) | antisolar arc | Figure 3-13 top |
| 10 | | *4 2 8 6* | ( like $\overline{5}$ ) | antisolar arc | |
| | 10 | *4 2 8 7 6 5 4 3 8 6* | | | Figure 3-13 bottom |
| 5 | | *4 2 6 7 8 3 4 6* | ( like $\overline{5}$ ) | antisolar arc | |
| | 3 | *4 2 6 7 8 3 4 6* | | | |
| | 2 | *4 2 6 7 8 3 4 5 6 7 8 3 4 6* - 2b | | | |
| 10 | | *3 7 1* | | | |
| 5 | | *3 8 7 5* | | | |

46 paths in 24 remaining types

1000 total

Refer to Figure 3-1 to identify the halos.

**Table 7.** Path types for the ray path figures.

| Halo | Ray Path Figure | Ray Path | Ray Path Type |
|---|---|---|---|
| Antisolar arc: | Figure 3-13 | *3 8 2 7* | *3 2 8 7* |
| | | *4 2 8 7 6 5 4 3 8 6* | *4 2 8 6* |
| Circumhorizon arc (plates): | Figure 6-13 | *3 2* | |
| Circumzenith arc (plates): | Figure 1-8 | *1 3* | |
| Circumzenith arc (Parry orientations): | Figure 3-9 | *3 1* | |
| Diffuse arcs: | Figure 2-26 | *3 5 6 7 1 3* | *3 1 5 6 7 3* |
| | | *3 1 7 5* | |
| Hastings arc (of upper suncave Parry): | Figure 3-14 | *3 2 7* | |
| Heliac arc: | Figure 3-12 | *5* | |
| | | *3 8 7* | |
| Infralateral arc: | Figure 2-14 | *1 3* | |
| Kern arc: | Figure 11-10 | *1 3 7* | |
| Lower tangent arc: | Figure 2-7 | *3 5* | |
| Lowitz arc (spinning plates): | Figure 11-4 | *4 6* | |
| Parhelic circle (plates): | Figure 1-17 | *3* | |
| | | *1 3 2* | |
| Parhelic circle (columns): | Figure 2-15 | *1* | |
| | | *3 2 6* | |
| Parhelion: | Figure 1-7 | *3 5* | |
| Parry arc, lower suncave: | Figure 6-17 | *4 6* | |
| Parry arc, lower sunvex: | Figure 6-14 | *5 7* | |
| Parry arc, upper suncave: | Figure 3-8 | *3 7* | |
| Parry arc, upper sunvex: | Figure 3-19 | *4 8* | |
| Parry infralateral arc: | Figure 3-11 | *1 7* | |
| Parry supralateral arc: | Figure 3-10 | *4 1* | |
| Subhelic arc: | Figure 2-23 | *2 3 5 1* | |
| Subparhelic circle (plates): | Figure 7-7 | *1 3 2 1* | *1 2 3 1* |
| | | *3 4 2 5* | *3 2 4 5* |
| | | *3 2 5 6 1 7 2 3* | *3 2 5 6 7 3* |
| Subparhelion: | Figure 7-4 | *3 2 5* | |
| Subsun (plates): | Figure 7-3 | *1* | |
| | | *3 2 6* | |
| Supralateral arc: | Figure 2-13 | *3 1* | |
| Tricker arc: | Figure 2-25 | *3 5 7 1 4* | *3 1 5 7 4* |
| | | *1 3 7 2 5 3 1* | *1 2 3 7 5 3 1* |
| Upper tangent arc: | Figure 2-7 | *3 7* | |
| Wegener arc: | Figure 2-21 | *3 2 7* | |
| 120° parhelion: | Figure 1-18 | *1 3 8 2* | |
| 9° halo: | Figure 10-14 | *3 1 6* | |
| 18.3° halo: | Figure 10-7 | *1 3 2 5* | |
| 23.8° halo: | Figure 10-20 | *3 2 5* | |

For each ray path figure in the text, the ray path is given as a sequence of numbered faces encountered by the ray. The path type is shown only when it differs from the path itself.

132

## APPENDIX F

## THE HALO SIMULATION PROGRAM

This appendix describes the computer program that created the halo simulations. The program was inspired by that of Pattloch and Tränkle [1984].

The program is a large loop, executed many times, consisting of the steps below. In one execution of the loop, the path of one ray from the sun is followed within one ice crystal until its intensity drops below 0.00001 of its original value. Except for total internal reflections, each encounter of the ray with a crystal face gives rise to an outgoing ray that lights a point on the celestial sphere. At that point in the simulation, a dot is plotted with probability equal to the intensity of the outgoing ray. (The simplified description in Chapter 1 ignored all but one of the outgoing rays.)

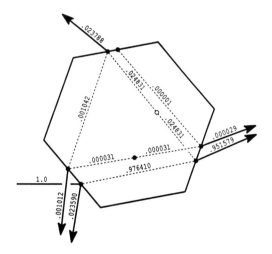

Path of a light ray within an oriented plate crystal, shown until the intensity has dropped below 0.00001. Here the two internal reflections from the basal faces happen to be total. At each remaining encounter with a crystal face there is an outgoing ray. The first outgoing ray here is a parhelic circle ray, the second is a parhelion ray, the third is a subparhelic circle ray, etc.

These are the instructions to the computer, but paraphrased for human consumption:

**1. Choose the incoming ray from the sun:** Choose a point at random on the sun's disk on the celestial sphere. The direction of the incoming ray to the crystal is opposite to that of the chosen point. (The position of the ray is chosen in step 4.)

Choose the wavelength of the incoming ray at random in the spectrum between green and red.

Set the four Stokes polarization parameters of the incoming ray equal to 1,0,0,0. The ray is therefore unpolarized and has intensity 1.

Associate with the incoming ray a beam cross section equal to 1.

**2. Choose the shape of the crystal:** Except for the pyramidal crystals of Chapter 10 and the triangular prisms of Figure 11-9, the crystal is a hexagonal prism. Choose its c/a ratio at random from a preassigned interval of numbers. If, for example, the interval ran from 0.2 to 0.4, then the crystals would be thick plates.

To approximate the imperfect symmetry of real crystals, vary the three distances between opposite prism faces randomly up to 2%, instead of making them exactly equal. Interfacial angles, of course, must remain constant at 120° or 90°.

**3. Choose the orientation of the crystal:** Suppose, for example, that the crystals are to be "columns with tilts of $x$ degrees." First orient the crystal so that it has two prism faces horizontal but is otherwise randomly oriented. For singly oriented columns, rotate the crystal randomly about its axis. Then tip the crystal axis so that it makes an angle with the horizontal plane. This angle — the tilt — is given by a normal distribution having mean zero and standard deviation $x$.

Thus, to say that the crystals have tilts of $x$, means not that the tilts are exactly $x$, but rather that their standard deviation is $x$. A table of values for the standard normal distribution shows that 68% of the crystals would be within $x$ of horizontal, and 95% within $2x$ of horizontal.

For Parry oriented columns the procedure is the same, except that the rotation angle — the amount of rotation about the crystal axis — is given by a normal distribution having mean zero and preassigned standard deviation, instead of being chosen randomly.

**4. Choose a point on the crystal for the incoming ray to hit:** Construct a disk with diameter equal to the longest diagonal of the crystal, and place it perpendicular to the incoming light and shading the entire crystal. Position the incoming ray so that it passes through a randomly chosen pinhole on the disk. This determines where, if at all, the incoming ray hits the crystal. If the ray misses the crystal, return to step 2. That is, choose another crystal shape, another crystal orientation, etc. (So crystal shape and orientation, but not size, can affect the chance that the crystal will be hit by the incoming ray.)

"Ray under consideration" will refer either to the incoming ray from the sun or to one of the ray segments within the crystal. At the present step, the ray under consideration is the incoming ray.

**5. Follow the ray under consideration until it meets a crystal face, and calculate the reflected and transmitted rays**, together with their Stokes parameters and beam cross sections. The intensities shown in the ray path diagrams in the text are products of the first Stokes parameter, the beam cross section, and the relevant refractive index.

One of the two reflected and transmitted rays is the outgoing ray from the crystal. The other is within the crystal and becomes the (next) ray under consideration.

**6. Plot a dot on the celestial sphere in the direction opposite to that of the outgoing ray:** The dot is the point lit by the outgoing ray. Plot it with probability equal to the intensity of the outgoing ray: choose a random number between zero and 1, and plot the dot if the intensity exceeds the chosen number. The density of dots in the simulation therefore indicates halo intensity.

**7(a). If the intensity of the ray within the crystal exceeds 0.00001 and if the ray has suffered fewer than 100 internal reflections, return to step 5.** That is, follow the ray path to the next crystal face and calculate new reflected and transmitted rays.

**(b). If not, return to step 1.** That is, choose another incoming ray, another crystal shape, another orientation, etc.

When the desired number of dots has been plotted, terminate or else repeat, using a different set of crystal shapes and orientations.

## APPENDIX G

## PARAMETERS OF THE HALO SIMULATIONS

This appendix lists parameter values used in making the simulations in the book. The parameters are:

**Sun elevation.**

**Number of crystals considered** (Column 1).

**Number of dots plotted** (Column 2). Many of the crystals considered by the computer are missed by the incoming ray, especially if the crystals are long columns or thin plates; see step 4 in Appendix F. Dots on the sun's disk or outside the region of sky covered by the simulation are not plotted. The number of dots is therefore less than the number of crystals considered.

**Mean c/a ratio** of the crystals (Column 3). See step 2 of Appendix F.

**Standard deviation of the crystal tilts**, in degrees (Column 4). A value of zero corresponds to perfectly horizontal crystals. See step 3 of Appendix F.

**Standard deviation of the rotation angle** (Column 5). Values of zero in both columns 4 and 5 correspond to perfect Parry orientations. See step 3 of Appendix F.

Do not be misled by the use of randomly oriented equidimensional crystals in the simulations. They are used to simulate the 22° halo, but as explained in Chapter 4, poorly oriented columns would often have done about as well.

| | Crystals considered | Dots plotted | Mean c/a | Standard deviation of tilts | Standard deviation of rotation angle |
|---|---|---|---|---|---|
| **Figure 1-6** sun elevation 23° | | | | | |
| oriented plates | 683,000 | 35,000 | 0.3 | 4.0 | |
| oriented plates | 251,000 | 20,000 | 0.2 | 40.0 | |
| **Figure 1-11** sun elevation 13° | | | | | |
| oriented plates | 1,356,000 | 20,000 | 0.2 | 1.5 | |
| **Figure 1-14** sun elevation 22.8° | | | | | |
| randomly oriented columns | 52,000 | 12,000 | 2.0 | | |
| oriented plates | 351,000 | 47,000 | 0.2 | 2.0 | |

In Figure 1-14 the basal faces of the plates, like those in the corresponding crystal sample, ranged from approximately equilateral triangles to approximately regular hexagons. These crystal shapes gave a relatively uniform intensity distribution along the parhelic circle. Compare with Figure C-3, where the basal faces were approximately regular hexagons.

| | Crystals considered | Dots plotted | Mean c/a | Standard deviation of tilts | Standard deviation of rotation angle |
|---|---|---|---|---|---|
| **Figure 2-4** | | | | | |
| sun elevation 19.7° | | | | | |
| singly oriented columns | 429,000 | 17,000 | 5.0 | 2.5 | |
| | | | | | |
| **Figure 2-9** | | | | | |
| sun elevation 20.9° | | | | | |
| singly oriented columns | 394,000 | 20,000 | 3.5 | 0.6 | |
| Parry oriented columns | 207,000 | 7,500 | 3.5 | 0.6 | 3.0 |
| oriented plates | 175,000 | 4,600 | 0.1 | 2.2 | |
| | | | | | |
| **Figure 2-12** | | | | | |
| sun elevation 22.9° | | | | | |
| singly oriented columns | 372,000 | 70,000 | 1.2 | 0.4 | |
| Parry oriented columns | 37,000 | 6,000 | 1.2 | 0.4 | 0.7 |
| randomly oriented crystals | 41,000 | 12,000 | 1.0 | | |
| oriented plates | 253,000 | 20,000 | 0.3 | 4.0 | |
| | | | | | |
| **Figure 2-18** | | | | | |
| sun elevation 21° | | | | | |
| singly oriented columns | 548,000 | 80,000 | 2.5 | 0.15 | |
| Parry oriented columns | 28,000 | 3,000 | 2.5 | 0.15 | 0.2 |
| randomly oriented crystals | 31,000 | 10,000 | 1.0 | | |
| oriented plates | 85,000 | 7,500 | 0.15 | 2.0 | |
| | | | | | |
| **Figure 3-1** | | | | | |
| sun elevation 20° | | | | | |
| Parry oriented columns | 329,000 | 50,000 | 2.0 | 0.15 | 0.15 |
| | | | | | |
| **Figure 3-3** | | | | | |
| sun elevation 20° | | | | | |
| singly oriented columns | 411,000 | 86,000 | 2.0 | 0.15 | |
| Parry oriented columns | 92,000 | 13,000 | 2.0 | 0.15 | 0.15 |
| randomly oriented crystals | 78,000 | 24,000 | 1.0 | | |
| oriented plates | 95,000 | 11,000 | 0.2 | 1.75 | |
| | | | | | |
| **Figure 3-5** | | | | | |
| sun elevation 20° | | | | | |
| singly oriented columns | 372,000 | 22,000 | 2.0 | 0.1 | |
| Parry oriented columns | 68,000 | 3,500 | 2.0 | 0.1 | 0.15 |
| randomly oriented crystals | 5,000 | 500 | 1.0 | | |
| oriented plates | 80,000 | 3,000 | 0.2 | 1.75 | |

In the preceding three simulations, each Parry oriented column had its alternate prism faces approximately equal but adjacent faces generally unequal; in each crystal the ratio between adjacent prism faces was chosen at random between 0.9 and 1.0. The effect on the simulations is to remove intensity variations along the antisolar arc that are not seen in the photographs. Compare Figure C-25.

| | Crystals considered | Dots plotted | Mean c/a | Standard deviation of tilts | Standard deviation of rotation angle |
|---|---|---|---|---|---|
| **Figure 3-22** | | | | | |
| sun elevation 13° | | | | | |
| singly oriented columns | 1,204,000 | 35,000 | 12.0 | 1.5 | |
| Parry oriented columns | 208,000 | 4,000 | 12.0 | 1.5 | 0.3 |

The Parry orientations in Figure 3-22 resemble teeter-totter oscillations. They weaken the suncave Parry arc but have little effect on the heliac arc.

| | Crystals considered | Dots plotted | Mean c/a | Standard deviation of tilts | Standard deviation of rotation angle |
|---|---|---|---|---|---|
| **Figure 3-26** sun elevation 10° | | | | | |
| singly oriented columns | 8,000 | 1,000 | 3.0 | 1.0 | |
| Parry oriented columns | 78,000 | 7,000 | 3.0 | 1.0 | 0.5 |
| randomly oriented crystals | 34,000 | 10,000 | 1.0 | | |
| oriented plates | 131,000 | 12,000 | 0.2 | 5.0 | |
| **Figure 4-3** sun elevation 20° | | | | | |
| singly oriented columns | 134,000 | 30,000 | 1.5 | 30.0 | |
| **Figure 6-7** sun elevation 38.5° | | | | | |
| oriented plates | 37,000 | 6,000 | 0.4 | 0.3 | |
| **Figure 6-10** sun elevation 72.4° | | | | | |
| randomly oriented crystals | 81,000 | 20,000 | 1.4 | | |
| oriented plates | 38,000 | 7,000 | 0.4 | 0.7 | |
| **Figure 6-11** sun elevation 72.4° | | | | | |
| singly oriented columns | 68,000 | 26,000 | 1.3 | 0.3 | |
| **Figure 6-12** sun elevation 72.4° | | | | | |
| singly oriented columns | 49,000 | 15,000 | 2.0 | 0.7 | |
| oriented plates | 39,000 | 7,000 | 0.4 | 0.7 | |
| **Figure 6-16** sun elevation 32° | | | | | |
| singly oriented columns | 121,000 | 8,000 | 2.0 | 0.1 | |
| Parry oriented columns | 114,000 | 4,000 | 2.0 | 0.1 | 1.0 |
| randomly oriented crystals | 121,000 | 6,000 | 1.0 | | |
| **Figure 6-19** sun elevation 58° | | | | | |
| singly oriented columns | 363,000 | 50,000 | 2.0 | 0.2 | |
| Parry oriented columns | 60,000 | 10,000 | 2.0 | 0.2 | 0.5 |
| randomly oriented crystals | 188,000 | 20,000 | 1.0 | | |
| **Figure 7-2** sun elevation 20° | | | | | |
| singly oriented columns | 84,000 | 8,000 | 2.0 | 0.8 | |
| oriented plates | 33,000 | 7,000 | 0.3 | 0.5 | |
| **Figure 7-6** sun elevation 20° | | | | | |
| Parry oriented columns | 450,000 | 5,000 | 2.0 | 0.8 | 8.0 |
| oriented plates | 809,000 | 3,400 | 0.3 | 0.5 | |
| **Figure 7-11** sun elevation 5° | | | | | |
| randomly oriented plates | 634,000 | 20,000 | 0.3 | | |
| oriented plates | 1,018,000 | 17,000 | 0.07 | 6.0 | |

| | Crystals considered | Dots plotted | Mean c/a | Standard deviation of tilts | Standard deviation of rotation angle |
|---|---|---|---|---|---|
| **Figure 10-5** | | | | | |
| sun elevation 14° | | | | | |
| randomly oriented pyramidal | | | | | |
| crystals as in Figure 10-7, left | 95,000 | 25,000 | | | |
| pyramidal crystals, Figure 10-7, right | 83,000 | 20,000 | | about 10.0 | |
| | | | | | |
| **Figure 10-12** | | | | | |
| sun elevation 70° | | | | | |
| randomly oriented pyramidal | | | | | |
| crystals as in Figure 10-10 | 151,000 | 60,000 | | | |
| | | | | | |
| **Figure 10-13** | | | | | |
| sun elevation 70° | | | | | |
| singly oriented columns | 152,000 | 40,000 | 2.0 | 10.0 | |
| pyramidal crystals as in Figure 10-10 | 91,000 | 40,000 | | 10.0 | |
| | | | | | |
| **Figure 10-16** | | | | | |
| sun elevation 55° | | | | | |
| singly oriented columns | 49,000 | 10,000 | 2.0 | 5.0 | |
| randomly oriented pyramidal | | | | | |
| crystals as in Figure 10-14 | 164,000 | 30,000 | | | |
| | | | | | |
| **Figure 10-18** | | | | | |
| sun elevation 32° | | | | | |
| pyramidal crystals as in Figure 10-20 | 79,000 | 20,000 | | 0.1 | |
| | | | | | |
| **Figure 10-19** | | | | | |
| sun elevation 32° | | | | | |
| singly oriented columns | 246,000 | 50,000 | 1.5 | 0.7 | |
| randomly oriented prisms | 116,000 | 30,000 | 1.0 | | |
| randomly oriented pyramidal | | | | | |
| crystals as in Figure 10-14 | 134,000 | 30,000 | | | |
| oriented plates | 101,000 | 7,000 | 0.3 | 4.0 | |
| pyramidal crystals as in Figure 10-20 | 77,000 | 20,000 | | 9.0 | |
| | | | | | |
| **Figure 11-3** | | | | | |
| sun elevation 25° | | | | | |
| singly oriented columns | 298,000 | 60,000 | 2.25 | 0.6 | |
| randomly oriented crystals | 42,000 | 15,000 | 1.0 | | |
| oriented plates | 240,000 | 10,000 | 0.05 | 1.5 | |
| | | | | | |
| **Figure 11-6** | | | | | |
| sun elevation 51° | | | | | |
| singly oriented columns | 99,000 | 30,000 | 2.0 | 0.2 | |
| randomly oriented crystals | 77,000 | 30,000 | 0.8 | | |
| oriented plates | 81,000 | 10,000 | 0.2 | 2.0 | |
| spinning plates | 181,000 | 30,000 | 0.2 | | |
| | | | | | |
| **Figure 11-7** | | | | | |
| sun elevation 51° | | | | | |
| singly oriented columns | 100,000 | 30,000 | 2.0 | 0.2 | |
| Parry oriented columns | 61,000 | 20,000 | 2.0 | 0.2 | 0.3 |
| randomly oriented crystals | 78,000 | 30,000 | 0.8 | | |
| oriented plates | 78,000 | 10,000 | 0.2 | 2.0 | |
| | | | | | |
| **Figure 11-9** | | | | | |
| sun elevation 14° | | | | | |
| oriented triangular plates | 65,000 | 20,000 | 0.4 | 0.5 | |

138

## FURTHER READING

Greenler [1980] is an excellent and readable general account of halos.

Minnaert [1954] is a delightful catalog of natural phenomena visible to the naked eye, including halos.

Visser [1942-1961] is a good reference for halos. Contains many halo observations, as well as theory.

Pernter and Exner [1922] is an early twentieth century classic on meteorological optics.

Wegener [1925] is an elegant and unified treatment of halos.

Tape [1980] presents a mathematical framework to treat various refraction halos.

Können [1985] describes small but striking effects of polarization in halos.

Pruppacher and Klett [1978] is a standard reference for cloud physics.

## REFERENCES

Besson, L., The Halos of November 1 and 2, 1913, *Monthly Weather Review*, *42*, 431-436, 1914.

Blake, J. R., Solar Halos in Antarctica, *Australian National Antarctic Research Expeditions Reports, Series A, 4*, 48 pp., Antarctic Division, Department of External Affairs, Melbourne, 1961.

Brand, W., and A. Wegener, Meteorologische Beobachtungen der Station Pustervig, *Meddelelser om Grønland, 42*, 553-560, 1912.

Bravais, M. A., Mémoire sur les Halos et les Phénomènes Optiques qui les Accompagnent, *Journal de l'école Royale Polytechnique, 31*, 1-270, 1847.

Drewry, D. J., and W. G. Rees, Observation of 120° Parhelia in Svalbard (letter), *Weather, 42*, 289, 1987.

Evans, W. F. J., and R. A. R. Tricker, Unusual Arcs in the Saskatoon Halo Display, *Weather, 27*, 234-238, 1972.

Fraser, A. B., What Size of Ice Crystals Causes the Halos?, *Journal of the Optical Society of America, 69*, 1112-1118, 1979.

Greenler, R., *Rainbows, Halos, and Glories*, 195 pp., Cambridge University Press, New York, 1980.

Greenler, R. G., and E. Tränkle, Anthelic Arcs from Airborne Ice Crystals, *Nature, 311*, 339-343, 1984.

Hakumäki, J., and M. Pekkola, Rare Vertically Elliptical Halos, *Weather, 44*, 466-473, 1989.

Hallett, J., Faceted Snow Crystals, *Journal of the Optical Society of America A, 4*, 581-588, 1987.

Hastings, C. S., *Light*, 224 pp., Charles Scribner's Sons, New York, 1902.

Hastings, C. S., A General Theory of Halos, *Monthly Weather Review, 48*, 322-330, 1920.

Hevelius, J., *Mercurius in Sole visus Gedani,...*, 173-176, Simon Reiniger, 1662.

Hobbs, P. V., *Ice Physics*, 837 pp., Oxford University Press, New York, 1974.

Humphreys, W. J., *Physics of the Air*, third edition, 676 pp., McGraw-Hill, New York, 1940. (Reprinted by Dover Publications, New York, 1964.)

Kobayashi, T., Vapour Growth of Ice Crystals Between –40 and –90 C, *Journal of the Meteorological Society of Japan, 43*, 359-367, 1965.

Kobayashi, T., and K. Higuchi, On the Pyramidal Faces of Ice Crystals, *Contributions from the Institute of Low Temperature Science A, 12*, 43-54, 1957.

Können, G. P., *Polarized Light in Nature*, 172 pp., Cambridge University Press, New York, 1985.

Liljequist, G. H., Halo-Phenomena and Ice-Crystals, in *Norwegian-British-Swedish Antarctic Expedition, 1949-52, Scientific Results*, Volume 2, Part 2A, 85 pp. and 11 plates, Norsk Polarinstitutt, Oslo, 1956.

Ling, C. S., Complex Solar Halo Observed at Ellendale, N. Dak., *Monthly Weather Review, 50*, 132-133, 1922.

Lowitz, T., Déscription d'un Météore Remarquable, Observé à St. Pétersbourg le 18 Juin 1790, *Nova Acta Academiae Scientiarum Imperialis Petropolitanae, 8*, 384-388, 1794.

Mariotte, E., *Oeuvres de Mr. Mariotte...,* Volume 1, *Traité des Couleurs,* pp. 272-281, Pierre Vander, 1717.

Mason, B., and L. G. Berry, *Elements of Mineralogy,* 550 pp., W. H. Freeman, San Francisco, 1968.

Minnaert, M., *The Nature of Light and Color in the Open Air,* 362 pp., Dover Publications, New York, 1954.

Moon, A. E., Parhelic Circle (letter), *Weather, 36,* 27, 1981.

Neiman, P. J., The Boulder, Colorado, Concentric Halo Display of 21 July 1986, *Bulletin of the American Meteorological Society, 70,* 258-264, 1989.

Parry, W. E., *Journal of a Voyage for the Discovery of a North-West Passage...,* 132 pp., John Murray, London, 1821. (Reprinted by Greenwood Press, New York, 1968.)

Pattloch, F., and E. Tränkle, Monte Carlo Simulation and Analysis of Halo Phenomena, *Journal of the Optical Society of America A, 1,* 520-526, 1984.

Pernter, J. M., and F. M. Exner, *Meteorologische Optik,* second edition, 907 pp., Wilhelm Braumüller, Vienna, 1922.

Pruppacher, H. R., and J. D. Klett, *Microphysics of Clouds and Precipitation,* 714 pp., D. Reidel, Boston, Mass., 1978.

Putnins, P., Der Bogen von Parry und andere unechte Berürungsbogen des gewöhnlichen Ringes, *Meteorologische Zeitschrift, 51,* 321-331, 1934.

Ripley, E. A., and B. Saugier, Photometeors at Saskatoon on 3 December 1970, *Weather, 26,* 150-157, 1971.

Steinmetz, H., and H. Weickmann, Zusammenhänge zwischen einer seltenen Haloerscheinung und der Gestalt der Eiskristalle, *Heidelberger Beiträge zur Mineralogie, 1,* 31-36, 1947.

Tape, W., Analytic Foundations of Halo Theory, *Journal of the Optical Society of America, 70,* 1175-1192, 1980.

Tränkle, E., and R. G. Greenler, Multiple-Scattering Effects in Halo Phenomena, *Journal of the Optical Society of America A, 4,* 591-599, 1987.

Tricker, R. A. R., *Introduction to Meteorological Optics,* 285 pp., Elsevier, New York, 1970.

Tricker, R. A. R., A Simple Theory of Certain Heliacal and Anthelic Halo Arcs..., *Quarterly Journal of the Royal Meteorological Society, 99,* 649-656, 1973.

Turner, F. M., and L. F. Radke, A Rare Observation of the 8° Halo, *Weather, 30,* 150-156, 1975.

Visser, S. W., *Handbuch der Geophysik,* Volume 8, Gebrüder Borntraeger, Berlin-Nikolassee, 1942-1961.

Wegener, A., Theorie der Haupthalos, *Aus dem Archiv der Deutschen Seewarte, 43,* 1-32, 1925.

Weickmann, H., *The Ice Phase in the Atmosphere,* Royal Aircraft Establishment, Library Translation No. 273, 95 pp., 42 figures, 50 plates, Ministry of Supply, London, 1948.

Young, Thomas, *A Course of Lectures on Natural Philosophy and the Mechanical Arts,* Volume 1, pp. 443-444, 786-787; Volume 2, pp. 303-308, printed for J. Johnson, London, 1807.

Zamorsky, A. D., Rare Cases of Halo and Rainbow (in Russian), *Priroda* No. 7, 81-85, 1959.